D1201406

Discarded

COLLEGE LIBRARY

Elementary

Thomas K. Maddox
Professor of Mathematics
Southeastern Louisiana College

Lawrence H. Davis
Associate Professor of Mathematics
Southeastern Louisiana College

Functions

Prentice-Hall, Inc., Englewood Cliffs, New Jersey

Elementary Functions

THOMAS K. MADDOX
LAWRENCE H. DAVIS

Englewood Cliffs, New Jersey

Current printing (last digit): 10 9 8 7 6 5 4 3 2 1

13-256727-X

Library of Congress Catalog Card Number 71-77848
Printed in the United States of America

PRENTICE-HALL INTERNATIONAL, INC., *London*
PRENTICE-HALL OF AUSTRALIA, PTY. LTD., *Sydney*
PRENTICE-HALL OF CANADA, LTD., *Toronto*
PRENTICE-HALL OF INDIA PRIVATE LTD., *New Delhi*
PRENTICE-HALL OF JAPAN, INC., *Tokyo*

Preface

The study of functions, a basic part of the education of college mathematics majors, is also rewarding to prospective teachers and to general readers. There are several reasons for this:

(1) The idea of a function has a broad appeal to students with widely different interests; it is readily recognized as an idea that can be applied in many different branches of knowledge. Functions that are commonly called "elementary" are useful in physics and chemistry; the same functions are also useful in economics, biology, and finance.

(2) The concept of a function is so basic to mathematics that it serves to unify substantial parts of elementary algebra, trigonometry, and analytic geometry.

(3) The study of functions motivates the development of manipulative skill in algebra.

(4) A good working knowledge of elementary functions gives the student a foundation for further work in mathematics and science.

This book can be read by a student who has had two years of standard high school mathematics. The reader is urged to work most of the exercises since they form an essential part of the book. A few of the exercises are marked with a dagger to indicate that they are more difficult than the rest.

We acknowledge our indebtedness to Mrs. David Lipscomb, who spent many hours in the preparation of the manuscript. We are especially grateful to Dr. Bruce Meserve for his invaluable editorial assistance.

Thomas K. Maddox
Lawrence H. Davis

Contents

Elementary Functions

1

Number Lines and Coordinate Planes

1-1 NUMBER LINES

A line is a *geometric* concept; the real numbers form a basic *algebraic* system. Geometry and algebra are brought together, and each is enriched, by using the set of real numbers to name the set of points on a line. A line with its points named by real numbers is called a **number line** or a **real line.**

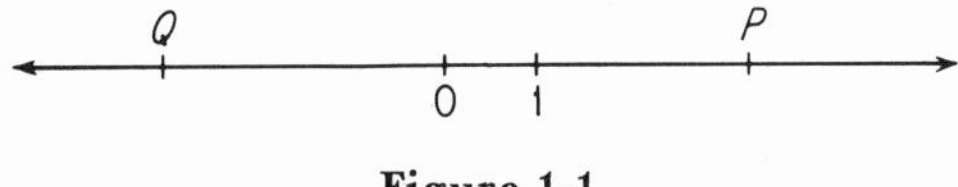

Figure 1-1

On a given line any two distinct points are chosen: one point is assigned the number 0 (and is called the **origin**); the other point is assigned the number 1 (and is called the **unit point**). The length of the line segment from point 0 to point 1 is taken to be 1. Then the other points along the line are assigned numbers in such a way that:

(1) If a point P is on the same side of the origin as the unit point and if the line segment from point 0 to P is x times as long as the line segment from point 0 to point 1, then P is assigned the number x.

(2) If a point Q is on the opposite side of the origin from the unit point and if the line segment from point 0 to Q is x times as long as the line segment from point 0 to point 1, then Q is assigned the number $-x$.

It is presumed that the reader has had some previous experience with numbers and is willing to start from this basic assumption: The points on a line *can* be numbered as described in the preceding paragraph; moreover, once the origin and the unit point have been selected, then to each point there corresponds a unique real number and to each real number there corresponds a unique point.

This is a "large" assumption to make. The system of real numbers is a complicated abstraction; it includes not only rational numbers such as 0, 2, and $-\frac{1}{2}$ but also irrational numbers such as $\sqrt{2}$, π, and $-\sqrt[3]{31}$. All of these real numbers, irrational as well as rational, are needed to name the points on a line.

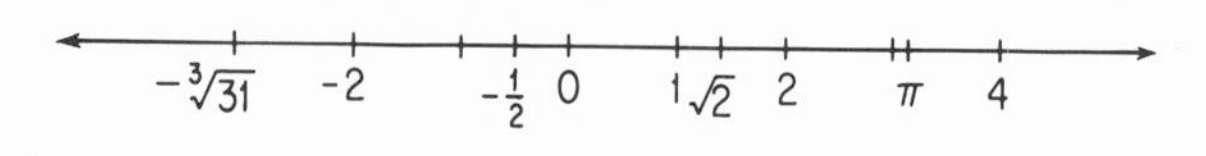

Figure 1-2

1-2 SOME SUBSETS OF A REAL NUMBER LINE

Each point of a number line is named by a real number that is the **coordinate** of the point. Each real number names a point on a number line. This point is called the **graph** of the number. The numbers (coordinates) and the points (graphs) are really different. However, because of the one-to-one correspondence between the points and the numbers, we will sometimes refer to "the point 6," for example, rather than use the longer expressions, "the point that is named 6" or "the point with coordinate 6."

Examples will now be given of a few important subsets of the real number system and their graphs.

The **integers** are the numbers in the infinite set $\{\ldots, -3, -2, -1, 0, 1, 2, 3, \ldots\}$. Some of the integral points along a number line are labeled in Figure 1-3. (Of course, there are far too many integral points for *all* of them to be labeled in a diagram.)

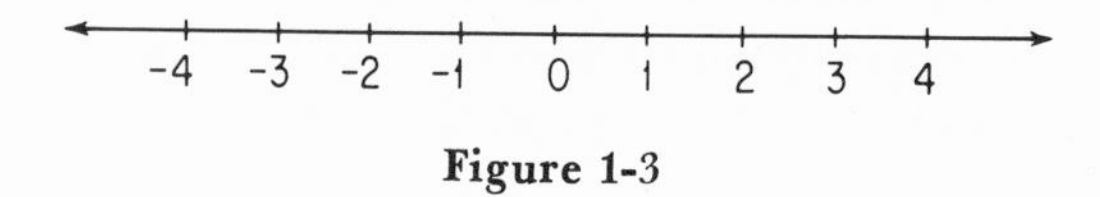

Figure 1-3

The **rational numbers** are the real numbers that can be written as the quotient of two integers, such as 5, $\frac{1}{2}$, 0, $-\frac{4}{5}$, and $\sqrt{16}$. It is useless to try to draw a diagram that will suggest the distribution along a real line of the points that have rational number coordinates. Between any two rational numbers there is another rational number; yet the set of rational numbers is not nearly sufficient to name *all* the points on a real line. In fact, between any two distinct points that have rational coordinates (no matter how close together they are) there is another point with a coordinate that is *not* rational.

The real numbers that are not rational are called **irrational numbers.** The numbers π and $\sqrt{2}$ are among the best-known examples. These numbers cannot be written in the form a/b, where a and b are integers. The reader may not have encountered many irrational numbers, but, in a sense, there are even more irrational than rational numbers. Fortunately, every irrational number can be approximated to any desired degree of accuracy by a rational number: for example, $\sqrt{2}$ is approximately 1.414 and π is approximately $\frac{22}{7}$.

The set of all **positive numbers** can be graphed on a number line as the set of points on the same side of the origin as the unit point. These points constitute the positive **half-line** (Figure 1-4).

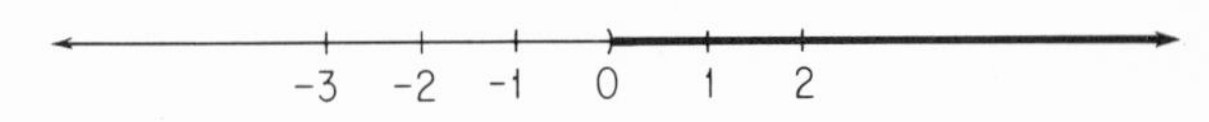

Figure 1-4 The positive half-line.

The set of negative numbers can be graphed on a number line as the set of points on the opposite side of the origin from the unit point. They, too, form a half-line (Figure 1-5).

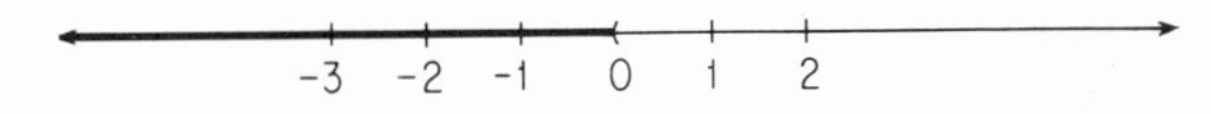

Figure 1-5 The negative half-line.

The number 0 is neither negative nor positive. Therefore, the set of **non-negative numbers** is different from the set of positive numbers; the former includes 0, but the latter does not. The inclusion of 0 is suggested in Figure 1-6 by the use of a square bracket at 0 instead of a single parenthesis.

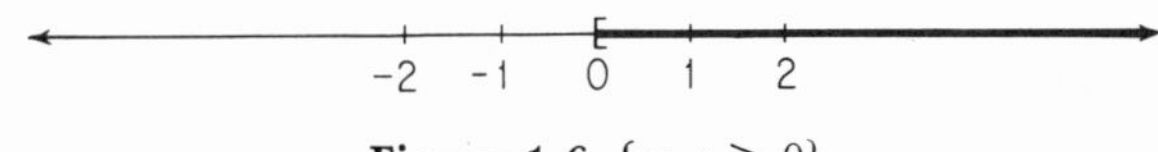

Figure 1-6 $\{x: x \geqq 0\}$

The set of all real numbers x such that x is between 2 and 5 is usually written $\{x: 2 < X < 5\}$. It is called an **open interval.** The graph of an open interval of numbers is an **open segment**—called "open" because it does not include its endpoints (Figure 1-7).

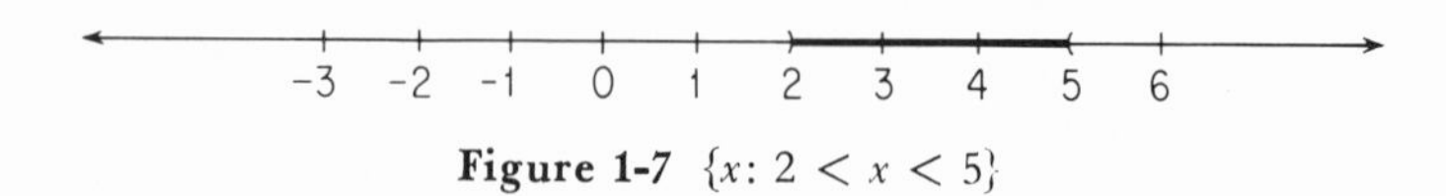

Figure 1-7 $\{x: 2 < x < 5\}$

The set $\{x: 2 \leqq X \leqq 5\}$ is an example of a **closed interval.** The graph of a closed interval is a **closed segment;** it contains both endpoints (Figure 1-8).

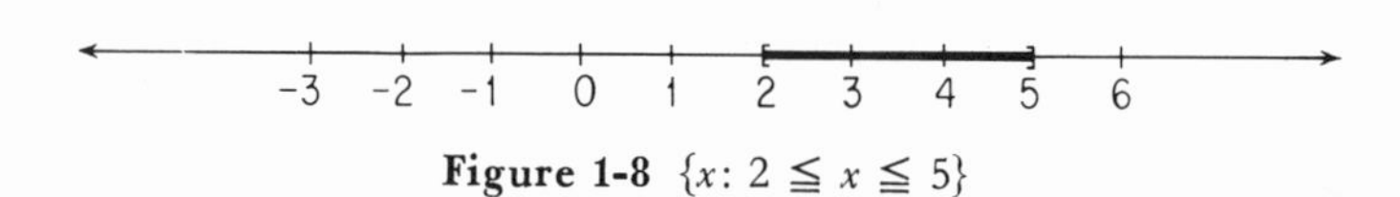

Figure 1-8 $\{x: 2 \leqq x \leqq 5\}$

An interval such as $\{x: 2 \leqq X < 5\}$ is a **half-open interval.** The interval includes the number 2 but not the number 5. As before, a square bracket is used in the diagram to indicate that an endpoint of a segment is included; a single parenthesis opening outward is used if the endpoint is not included (Figure 1-9).

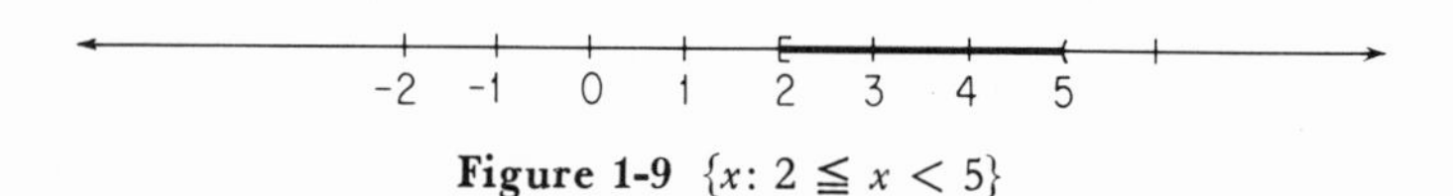

Figure 1-9 $\{x: 2 \leqq x < 5\}$

1-1 and 1-2 EXERCISES

Use a colored pen or pencil as an aid in graphing each of these intervals on a real line. In some of the exercises you may wish first to formulate a simpler description of the set than the one that is given.

1. $\{x: -2 < x < 4\}$

2. $\{x: x > 5\}$

3. $\{x: x \leqq 2\}$

4. $\{x: -3 \leqq x \leqq 0\}$

5. $\{x: x > -1\}$

6. $\{x: 2x \leqq 5\}$

7. $\{x: 2x + 1 \leqq 5\}$

8. $\{x: 2x + 1 < 3x\}$

9. $\{x: \frac{x + 2}{3} < 4\}$

10. $\{x: x^2 \leqq 4\}$

11. $\{x: x^2 \leqq 2\}$

12. $\{x: (x + 2)(x - 3) < 0\}$

1-3 THE ABSOLUTE VALUE OF A NUMBER

Two real numbers that differ only in sign, such as 5 and -5, are said to have the same absolute value. The positive number 5 is the common absolute value of both 5 and -5; 7 is the common absolute value of both 7 and -7.

Definition: The **absolute value** of a real number x is
1. x if x is positive or zero.
2. $-x$ if x is negative.

The absolute value of x is denoted by $|x|$. Thus, $|5| = 5$; also $|-7| = -(-7) = 7$. Four observations should be made immediately:

(1) A number and its negative have the same absolute value.

(2) The absolute value of a number x is equal to the absolute value of a number y if and only if either x and y are the same number or each of the numbers is the negative of the other.

(3) The number 0 is the only number whose absolute value is 0.

(4) A number is different from 0 if and only if its absolute value is a positive number.

These facts, which will be used many times in this and later chapters, can be stated more compactly:

(1) $|x| = |-x|$ for all numbers x.
(2) $|x| = |y| \Leftrightarrow x = y$ or $x = -y$.
(3) $|x| = 0 \Leftrightarrow x = 0$.
(4) $x \neq 0 \Leftrightarrow |x| > 0$.

The symbol $\Leftrightarrow$ means "is logically equivalent to." It is also read "if and only if." The statement

$$|x| = |y| \Leftrightarrow x = y \quad \text{or} \quad x = -y$$

makes the assertion that the simple sentence

$$|x| = |y|$$

is logically the same as the compound sentence

$$x = y \quad \text{or} \quad x = -y.$$

Every time the symbol $\Leftrightarrow$ is used, a proposition and its converse are both asserted. For example,

$$|x| = |y| \Leftrightarrow x = y \quad \text{or} \quad x = -y$$

might be restated in two parts as follows:

(1) If $|x| = |y|$, then $x = y$ or $x = -y$.
(2) If $x = y$ or $x = -y$, then $|x| = |y|$.

A mathematical argument or proof may proceed by moving from one sentence to another that is the logical equivalent of the first sentence, then on to a third sentence that is the logical equivalent of the second, etc. Thus a "chain" of logically equivalent sentences may be constructed. For example, suppose we are asked to solve the equation

$$|x - 4| = 6.$$

We construct a brief chain of reasoning:

$$|x - 4| = 6 \Leftrightarrow x - 4 = 6 \quad \text{or} \quad x - 4 = -6$$
$$\Leftrightarrow x = 10 \quad \text{or} \quad x = -2.$$

The conclusion is that 10 and -2 satisfy the given equation and are the only numbers that do.

The reader should also understand that the chain of reasoning in the previous example asserts that the sentence

$$|x - 4| = 6$$

is logically the same as the sentence

$$x - 4 = 6 \quad \text{or} \quad x - 4 = -6$$

which is logically the same as the sentence

$$x = 10 \quad \text{or} \quad x = -2.$$

1-3 EXERCISES

1. Solve each of these equations:

(a) $|t| = 8$ (b) $|s - 4| = 10$
(c) $|x + 4| = 6$ (d) $|y + 7| = 3$
(e) $|2x - 1| = 6$ (f) $2|x - 1| = 6$
(g) $|x + 2| + 1 = 3$ (h) $|x - 2| = -3$

2. One of the following generalizations is valid, the other is not:

(a) $|a + b| = |a| + |b|$ for all numbers a and b.
(b) $|ab| = |a| \cdot |b|$ for all numbers a and b.

Which is the valid generalization? Give reasons. Use an inequality symbol to modify the invalid generalization so as to obtain a new generalization that will be correct and meaningful.

3. Explain why $|x| = \sqrt{x^2}$ for every real number x.

1-4 THE DISTANCE BETWEEN TWO POINTS ON A NUMBER LINE

Distance between points is a basic idea in mathematics, geography, and physics. In mathematics, distance is simply a non-negative real number, a measure with no units of measurement specified. The absolute value symbol is useful in defining the distance between two points on a number line.

Definition: Suppose points A and B on a number line have respective coordinates a and b. Then the **distance between A and B** is $|a - b|$. In brief notation, $d(A, B) = |a - b|$, where $d(A, B)$ denotes "the distance between points A and B."

For example, if 6 is the coordinate of A and -4 is the coordinate of B, then $d(A, B) = |6 - (-4)| = 10$ (Figure 1-10).

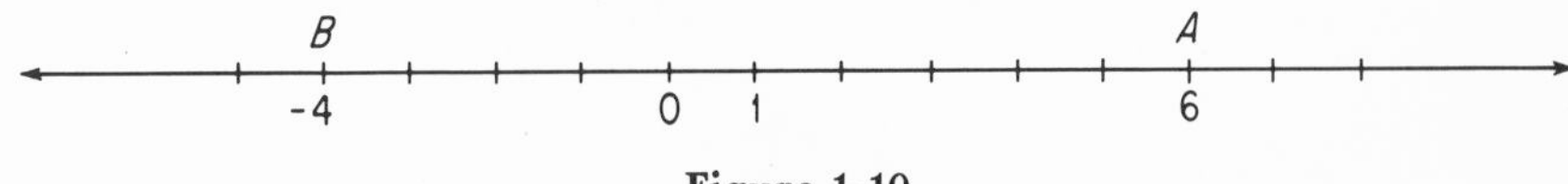

Figure 1-10

If A, B, and C denote arbitrary points on a number line, then these four generalizations are valid:

(1) $d(A, B) \geqq 0$.
(2) $d(A, B) = 0 \Leftrightarrow A$ and B are the same point.
(3) $d(A, B) = d(B, A)$.
(4) $d(A, B) \leqq d(A, C) + d(C, B)$.

The student is asked to justify these generalizations in Exercise 4 at the end of this section.

Note that the distance between the origin and the point with coordinate x is $|0 - x| = |-x| = |x|$. Two points are equidistant from the origin if and only if their coordinates have the same absolute value.

We have previously solved the equation $|x - 4| = 6$ and found the two solutions 10 and -2. These solutions have a useful interpretation in terms of distance. Since $|x - 4|$ gives the distance between the two points on a number line with coordinates x and 4, solutions to $|x - 4| = 6$ are coordinates of the two points that are each 6 units from the point with coordinate 4 (Figure 1-11).

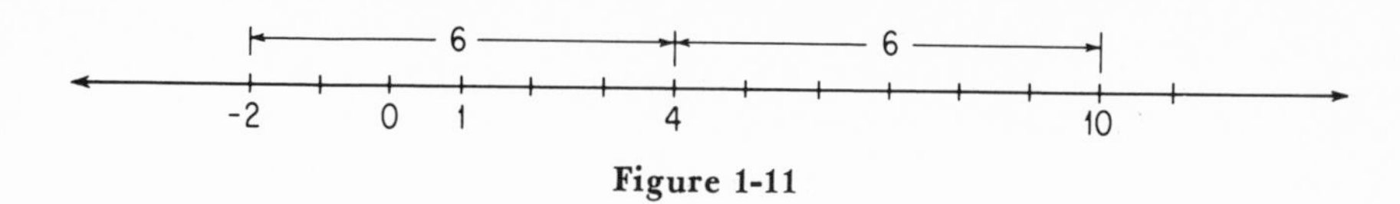

Figure 1-11

Without using pencil and paper we can solve the equation $|x - 1| = 8$ by visualizing a number line and asking: What points on the line are 8 units distant from point 1? There are two of these points, and their coordinates are -7 and 9. Therefore -7 and 9 are the only two solutions to the equation $|x - 1| = 8$.

Similarly, we can solve the equation $|x + 3| = 5$ by thinking of the distance interpretation. Observe first that $|x + 3|$ is the same as $|x - (-3)|$, so asking for the solution of $|x + 3| = 5$ is the same as asking for the coordinates of the two points that are 5 units from the point with coordinate -3. These coordinates are -8 and 2 (Figure 1-12).

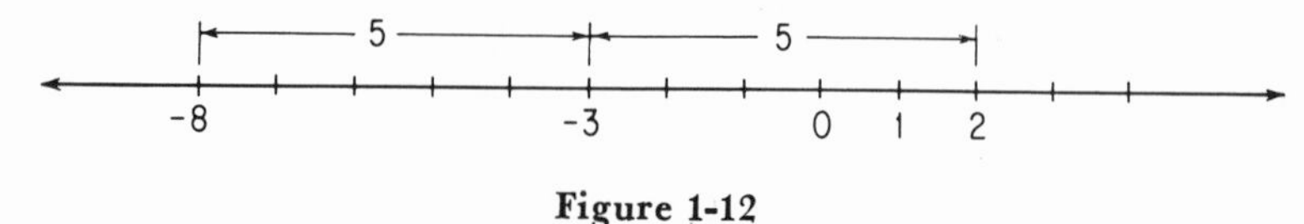

Figure 1-12

The absolute value symbol is useful in describing intervals of real numbers. Some examples follow. A distance interpretation will help the reader to see just what set of points is described in each example.

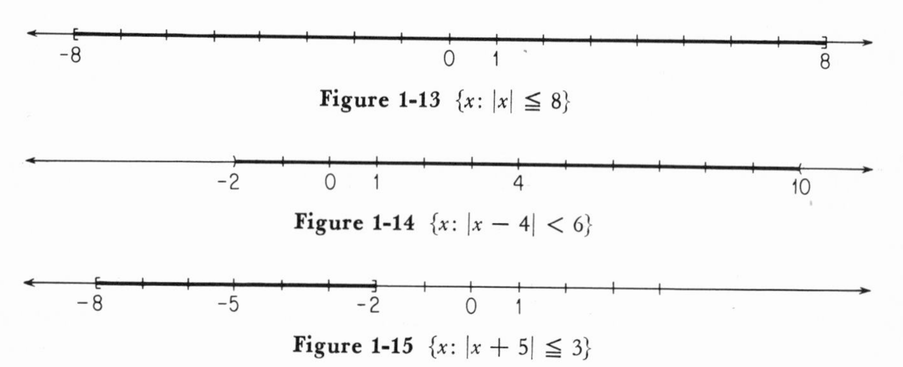

Figure 1-13 $\{x: |x| \leqq 8\}$

Figure 1-14 $\{x: |x - 4| < 6\}$

Figure 1-15 $\{x: |x + 5| \leqq 3\}$

1-4 EXERCISES

1. Solve each of these equations (preferably without using pencil and paper) by thinking of a distance interpretation:

(a) $|x - 2| = 10$ (b) $|x + 6| = 5$
(c) $|x - 3| = 0$ (d) $|x - 4| = -2$
(e) $|t - 10| = 2$ (f) $|t + 10| = 2$
(g) $2|s - 6| = 4$ (h) $|x - 3| = |x - 9|$

2. Graph each of these sets of real numbers:
(a) $\{x: |x| \leqq 2\}$ (b) $\{x: |x| > 1\}$
(c) $\{x: |x - 2| \leqq 3\}$ (d) $\{x: |x + 2| \leqq 3\}$
(e) $\{x: |x + 1| < 1\}$ (f) $\{x: 3|x| \leqq 12\}$
(g) $\{x: |x - 3| \leqq 0\}$ (h) $\{x: 4 \leqq |x - 1|\}$

3. Find the coordinate of the point that is
(a) Halfway between the points with coordinates -2 and 10.
(b) Halfway between the points with coordinates -8 and 15.
(c) Halfway between the points with coordinates c and d (where $c \neq d$).

4. Justify the four generalizations (made in this section) about distance between points on a number line.

5. On a number line let A denote the point with coordinate -2 and B

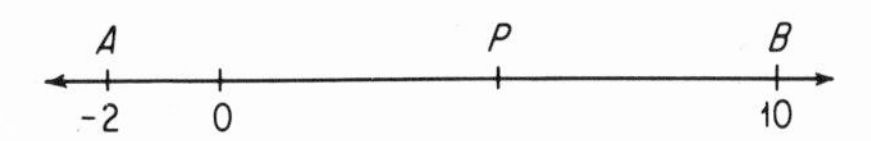

denote the point with coordinate 10. Find the coordinate of a point P between A and B such that:
(a) $d(A, P) = 3 \cdot d(P, B)$
(b) $d(A, P) = \frac{1}{2} \cdot d(P, B)$
(c) $d(A, P) = 10 \cdot d(P, B)$
(d) $d(A, P) = q \cdot d(P, B)$, where q is an arbitrary positive number.

6. An object moves along a number line in such a way that the coordinate of its position P at time t can be calculated by this rule: $P(t) = 5 - 2t$.
(a) What is the distance between the object and the origin when $t = 0$? When $t = 10$?
(b) Write the rule that expresses the distance between the object and the origin at time t. (Note that this is different from the "position" rule.)
(c) Write the rule that expresses the distance between the object and the point with coordinate 2 at time t.

7. Two objects move along the same number line. The position of the first one is given by the rule of Exercise 6. The coordinate of the position P' of the second object is given by the rule $P'(t) = t^2 + 5$.
(a) What is the distance between the objects when $t = 10$?

(b) Write the rule that expresses the distance between the objects at an arbitrary time t.

1-5 COORDINATE PLANES

A single number names a point on a number line, but a pair of numbers is required to name a point in a given plane. This agrees with our intuitive understanding of a line as being "one-dimensional" and a plane as being "two-dimensional."

A **rectangular coordinate system** is the most widely used scheme for naming points in a plane. Choose two perpendicular lines in the plane with one line called the ***x*-axis** and the other line called the ***y*-axis.** Their point of intersection is taken as the origin of each axis (Figure 1-16). With each point on the x-axis associate a unique real number (as in Section 1-1); similarly, with each point on the y-axis associate a unique real number. (For some applications, unit length on the x-axis must be the same as unit length on the y-axis. But this is not a general requirement for a coordinate plane; for many applications the unit lengths on the two axes may be different.)

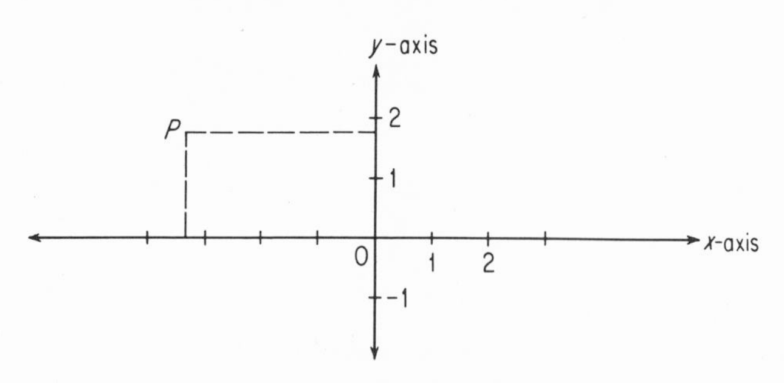

Figure 1-16

Let P denote an arbitrary point in the plane. There is a unique line through P perpendicular to the x-axis. Suppose it intersects that axis in a point with coordinate x. Also there is a unique line through P perpendicular to the y-axis. Suppose it intersects the y-axis in a point with coordinate y. Then the point P is denoted by the ordered pair of numbers (x, y).

We make the following basic assumption: The points of a plane can be named in the manner just described. Moreover, once the points on each of the axes have been assigned coordinates, then each point of the plane is named by a unique ordered pair of real numbers, and to each ordered pair of numbers there corresponds a unique point in the plane.

Suppose point P is denoted by the ordered pair (x, y). These two numbers are called the **coordinates** of P: the first number of the pair is called the **x-coordinate** (or **abscissa**), and the second number of the pair is called the **y-coordinate** (or **ordinate**).

The four quadrants are named in Figure 1-17. A point is in Quadrant I if and only if both coordinates are positive; a point is in Quadrant II if and only if the x-coordinate is negative and the y-coordinate is positive; a point is in Quadrant III if and only if both coordinates are negative; and a point is in Quadrant IV if and only if the x-coordinate is positive and the y-coordinate is negative (Figure 1-17).

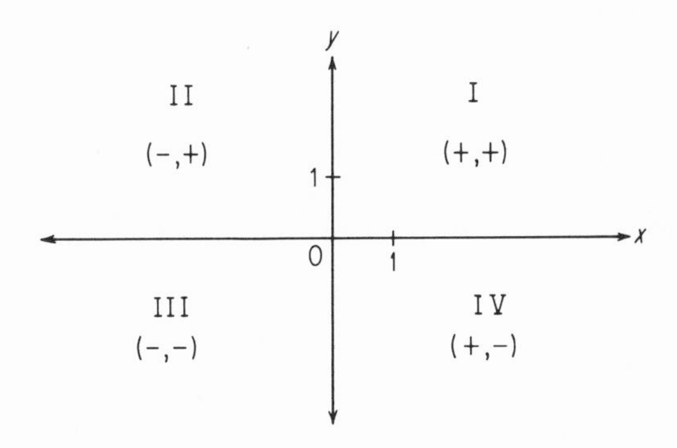

Figure 1-17

The points along the axes do not belong to any of the four quadrants. In fact, a point is on the x-axis if and only if its y-coordinate is 0; that is, its coordinates are $(k, 0)$ for some k. A point is on the y-axis if and only if its x-coordinate is 0; that is, its coordinates are $(0, k)$ for some k.

1-6 SOME SUBSETS OF A PLANE

It is easy enough to catch on to the scheme just described for naming individual points in a plane. But the reader needs to do more than this—he needs to have a good understanding of the description of various subsets of a plane. Some of these subsets are "regions"; some

some are "curves." Some are "connected"; some are "disconnected." Subsets of a plane come in great variety.

The usual notation for describing subsets of a number line (Section 1-3) is of the form $\{x: |x + 1| = 2\}$, where x is to satisfy the specified condition. A similar notation is used for subsets of a coordinate plane, but the reader must remember that a point in a plane has *two* coordinates. A given condition may involve only one of these coordinates, or it may involve both. Consider, for example, the graph of $\{(x, y): x = 2\}$; that is, all those points with coordinates (x, y) for which the x-coordinate is 2. This condition imposes no requirement whatsoever on the y-coordinate of the point; a point P belongs to the graph if and only if its x-coordinate is 2. The graph of this subset is shown in Figure 1-18. The graph is a straight line parallel to the y-axis and situated 2 units to the right of the y-axis. The coordinates of each point (x, y) on this line satisfy the condition $x = 2$; the coordinates of each point (x, y) that is not on this line fail to satisfy the condition $x = 2$.

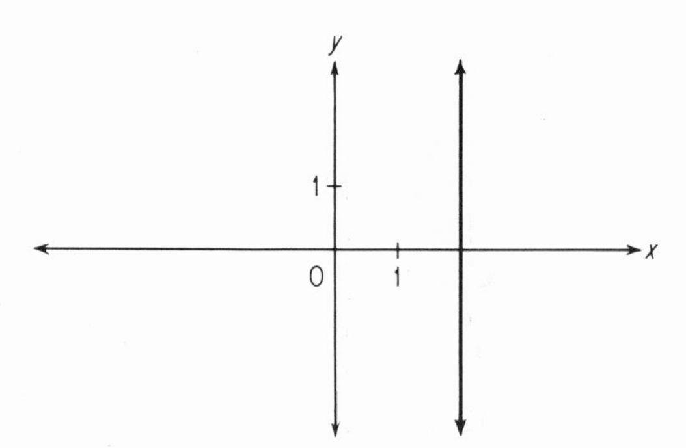

Figure 1-18 $\{(x, y): x = 2\}$

As a second example, consider the subset $\{(x, y): xy \geqq 0\}$. This condition involves both coordinates; the product of the two must not be negative. The coordinates of the points in Quadrants II and IV do not satisfy the condition $xy \geqq 0$. All other points (x, y) of the plane have coordinates such that $xy \geqq 0$. When we graph $\{(x, y): xy \geqq 0\}$ we follow the convention of shading or hatching the appropriate region of the plane (Figure 1-19).

As a third example, consider $\{(x, y): |x| < 2 \text{ and } |y| < 3\}$. The

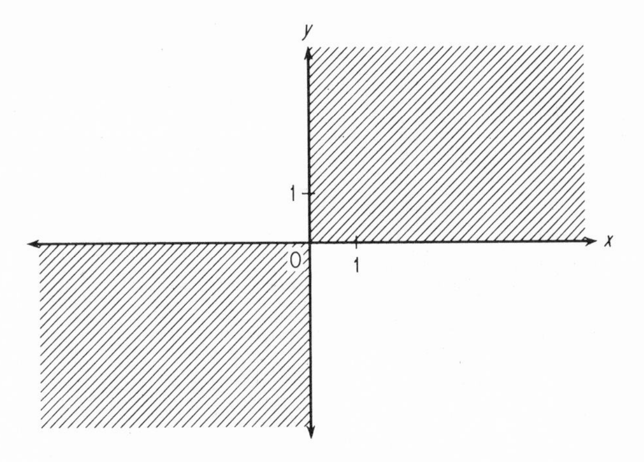

Figure 1-19 $\{(x, y)\colon xy \geqq 0\}$

graph of this set consists of the interior points of a rectangular region (Figure 1-20). The boundary of the region is dashed to indicate that the points of the rectangle are not included in the graph.

Suppose in the preceding example that we change the connective "and" in the description of the set to "or." This small change in wording makes a big difference in meaning. The graph $\{(x, y)\colon |x| < 2 \text{ or } |y| < 3\}$ includes every point (x, y) such that $|x| < 2$; the graph also includes every point such that $|y| < 3$. The graph consists of 2 infinite "bands" (Figure 1-21).

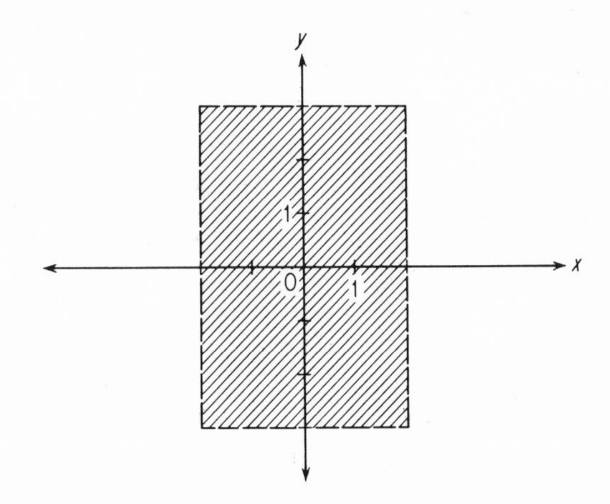

Figure 1-20 $\{(x, y)\colon |x| < 2 \text{ and } |y| < 3\}$

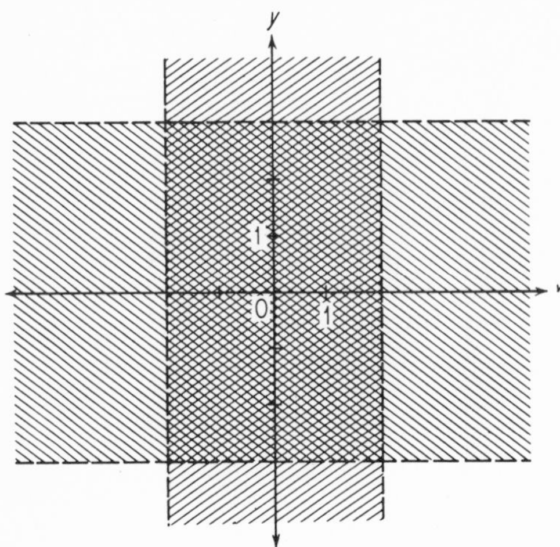

Figure 1-21 $\{(x, y): |x| < 2 \text{ or } |y| < 3\}$

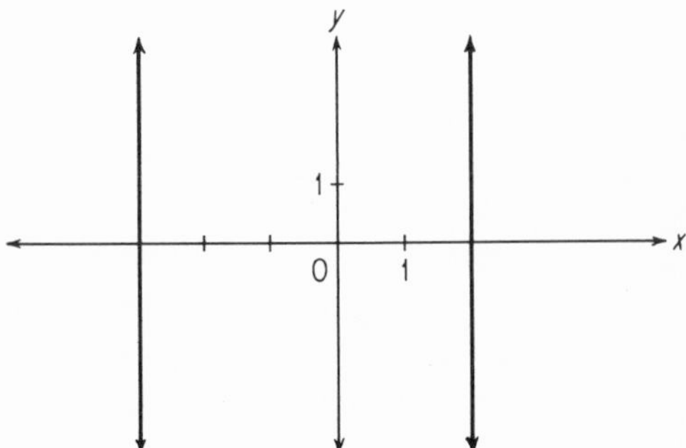

Figure 1-22 $\{(x, y): (x - 2)(x + 3) = 0\}$

As a final example, consider $\{(x, y): (x - 2)(x + 3) = 0\}$. One of the important properties of the real number system is that the product of two numbers is zero if and only if one of the numbers is zero. Applying this property, we see that

$$(x - 2)(x + 3) = 0 \Leftrightarrow x = 2 \quad \text{or} \quad x = -3.$$

Therefore,

$$\{(x, y): (x - 2)(x + 3) = 0\} = \{(x, y): x = 2 \quad \text{or} \quad x = -3\}.$$

The graph of this set consists of two parallel lines (Figure 1-22).

1-5 and 1-6 EXERCISES

Graph on a coordinate plane:

1. $\{(x, y): xy \leqq 0\}$

2. $\{(x, y): x \leqq 5\}$

3. $\{(x, y): |x| \leqq 5\}$

4. $\{(x, y): |y| > 2\}$

5. $\{(x, y): |x - 2| < 4\}$

6. $\{(x, y): (y - 5)(y + 1) = 0\}$

7. $\{(x, y): (x - 3)(x - 2)(y - 4) = 0\}$

8. $\{(x, y): x \text{ is an integer}\}$

9. $\{(x, y): y^2 = 9\}$

10. $\{(x, y): x \text{ is an integer and } y \text{ is an integer}\}$

11. $\{(x, y): x^2 = 4\}$

12. $\{(x, y): x \text{ is an integer or } y \text{ is an integer}\}$

1-7 THE LINE DETERMINED BY TWO POINTS IN A COORDINATE PLANE

What is a line? We have used the word freely, as if there could be no doubt about its meaning. Yet the reader, who perhaps never has questioned his understanding of "line," may not find it easy to answer this: Is the point (10, 33) on the line that contains the points (2, 1) and (3, 5)? (See Figure 1-23.)

It helps some, but not much, to plot all three points and then place a straightedge alongside, revealing that (10, 33) is either on the line or close to it. But, at best, the straightedge procedure can give only an approximate answer. We want more than this—after all, a mathematical "line" has no thickness, so (10, 33) might be just off the line. Precisely what points does the line consist of? A definition is needed.

We will avoid the deeper questions that are involved, give a straightforward definition, and then a brief explanation. The student

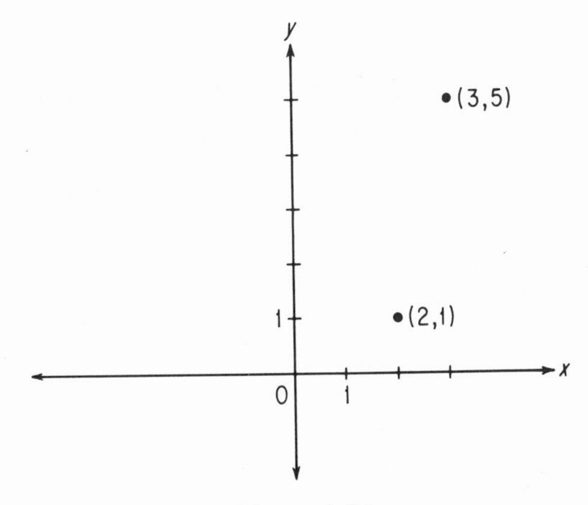

Figure 1-23

who knows something about similar triangles (from high school geometry) can use that knowledge in justifying the definition.

Definition: The line determined by the points (2, 1) and (3, 5) is the graph of $\{(x, y): y - 1 = 4(x - 2)\}$.

Except for the point (2, 1), the points that satisfy

$$y - 1 = 4(x - 2)$$

are also the points that satisfy the equation

$$\frac{y - 1}{x - 2} = 4.$$

The ratio form of the equation reveals the motivation for the definition. This ratio, which may be thought of as rise/run (Figure 1-24),

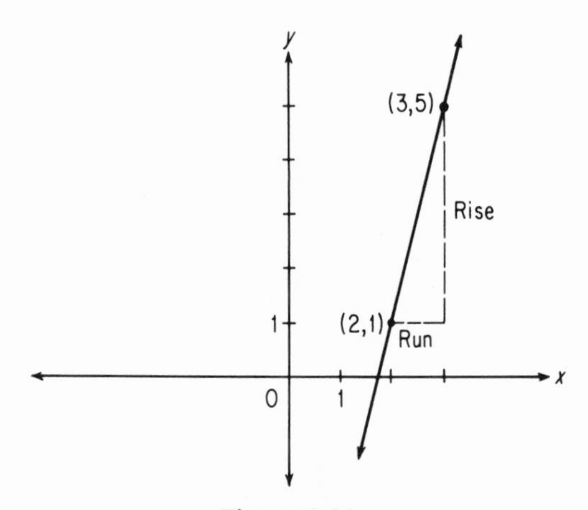

Figure 1-24

is called the **slope** of the line. It is required to be 4 in this case because for the two points that are given,

$$\frac{\text{rise}}{\text{run}} = \frac{5 - 1}{3 - 2}.$$

The ratio rise/run must be held constant for all pairs of points on the line.

To answer our original question as to whether (10, 33) is on the line, we can quickly check whether or not

$$\frac{33 - 1}{10 - 2} = 4.$$

Since the coordinates of the point (10, 33) satisfy the equation of the line, the point is on the line. Notice that we could just as logically have verified that (10, 33) satisfies

$$y - 1 = 4(x - 2).$$

The slope of the line determined by the points (2, 1) and (3, 5) is 4, but other lines have other slopes, and lines that are parallel to the y-axis do not have a slope. The general definition of "the line determined by two points" must include both lines that have a slope and lines that do not have slope.

Definition: The line determined by two distinct points (a, b) and (c, d) is the graph of the set:
1. $\{(x, y): x = a\}$ if $a = c$; that is, if the two points have the same x-coordinate.
2. $\{(x, y): y - b = m(x - a)\}$ if $a \neq c$, where m (called the **slope**) is defined to be

$$\frac{d - b}{c - a}.$$

A few examples should help to make this definition clear.

EXAMPLE 1: The line determined by the points (1, 3) and (1, 7) is the graph of $\{(x, y): x = 1\}$. This line is parallel to the y-axis. It has no slope (Figure 1-25).

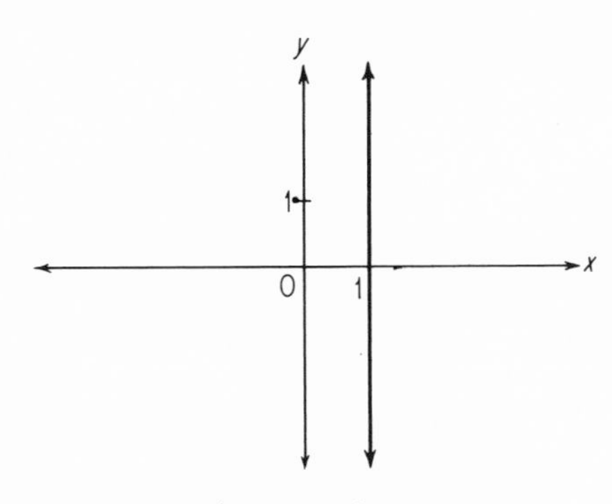

Figure 1-25

EXAMPLE 2: The line determined by the points (1, 4) and (6, 14) has slope

$$\frac{14 - 4}{6 - 1} = 2.$$

The line (Figure 1-26) is the graph of

$$\{(x, y): y - 4 = 2(x - 1)\}.$$

The line is also the graph of

$$\{(x, y): y - 14 = 2(x - 6)\}.$$

The reader can verify that

$$y - 4 = 2(x - 1) \Leftrightarrow y - 14 = 2(x - 6).$$

It makes no difference which one of the two given points is thought of as the (a, b) of the definition. (See Exercise 6.)

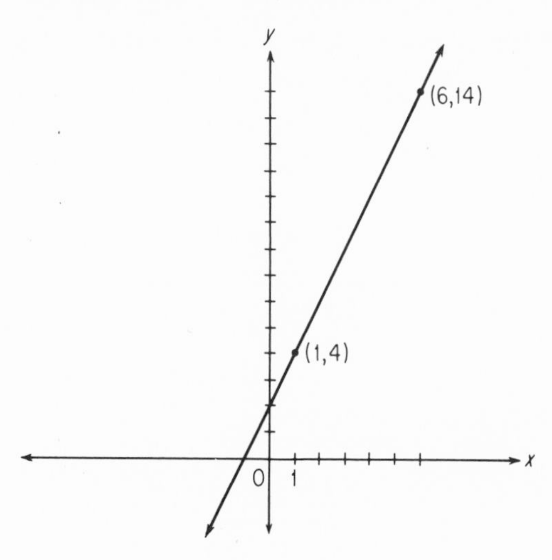

Figure 1-26

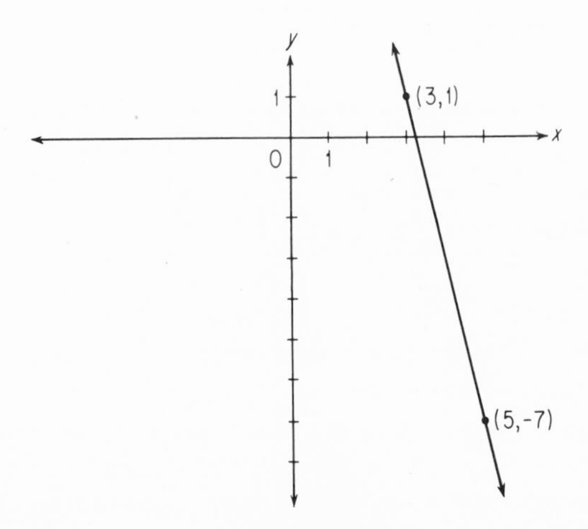

Figure 1-27

EXAMPLE 3: The line determined by the points (3, 1) and (5, −7) has slope −4, since

$$\frac{-7 - 1}{5 - 3} = -4.$$

The line is the graph of

$$\{(x, y) : y - 1 = -4(x - 3)\}.$$

(See Figure 1-27.)

EXAMPLE 4: The line determined by the points (1, 4) and (7, 4) is the graph of $\{(x, y) : y = 4\}$. For this line the slope is 0, since

$$\frac{4 - 4}{7 - 1} = 0.$$

(See Figure 1-28.)

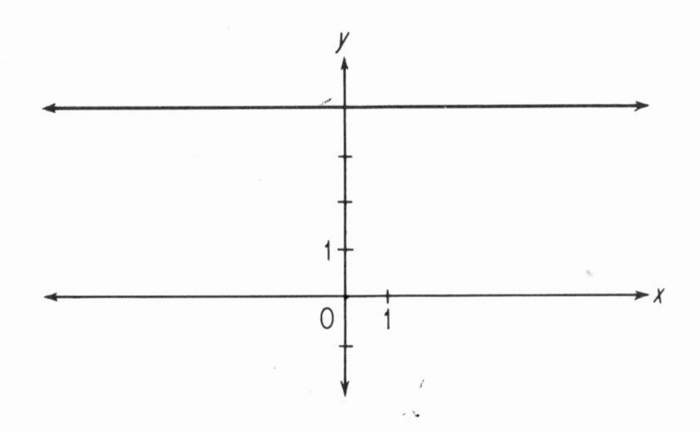

Figure 1-28

EXAMPLE 5: The line that contains the point (−4, 6) and has slope 7 is the graph of

$$\{(x, y) : y - 6 = 7(x + 4)\}.$$

Note that a line that has a slope can be specified by giving one of its points and its slope. (See Figure 1-29.)

Slope is clearly an important characteristic of any line that has slope. The slope of a line gives a measure of how rapidly the line is "rising" (from left to right), or perhaps "falling." A line has positive slope if and only if it "rises" from left to right; a line has negative slope if and only if it "falls" from left to right; a line has 0 slope if and only if it is parallel to the x-axis; it has no slope (its slope is undefined) if and only if the line is parallel to the y-axis.

These results suggest a question: If we let A, B, and C stand for arbitrary numbers (except that not both A and B are 0), is the graph of the set $\{(x, y): Ax + By = C\}$ a line? The answer is yes. To verify this, consider two cases:

(1) Suppose $B = 0$. Then $A \neq 0$. The equation $Ax + 0y = C$ is equivalent to $x = -C/A$. This is the equation of a line parallel to the y-axis.

(2) Suppose $B \neq 0$. Then $Ax + By = C$ is equivalent to $y - C/B = (-A/B)x$. According to Section 1-7, this equation defines a line containing the point $(0, C/B)$ and having slope $-A/B$.

EXAMPLE 1: Sketch the graph of $\{(x, y): 2x + 3y = 6\}$. From our general discussion we know that the graph is a line; so select

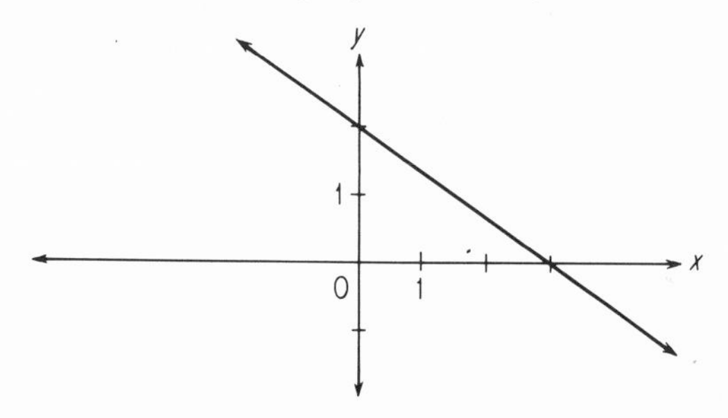

Figure 1-30 $\{(x, y): 2x + 3y = 6\}$

any two points in the set $\{(x, y): 2x + 3y = 6\}$, plot these two points, and then draw the line that contains these points. The intercepts (points where the line crosses the axes) are usually the easiest points to determine. For this line the intercepts are $(0, 2)$ and $(3, 0)$. Frequently a third point or the slope is used as a check on the graph. The slope $-A/B$ of this line is $-\frac{2}{3}$ (Figure 1-30).

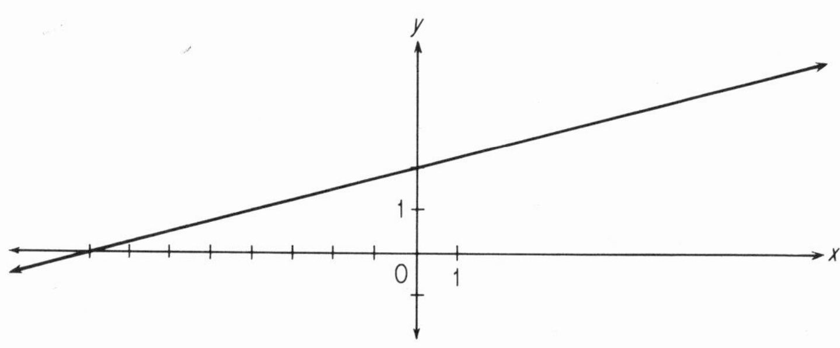

Figure 1-31 $\{(x, y): x - 4y = -8\}$

EXAMPLE 2: Sketch the graph of $\{(x, y): x - 4y = -8\}$. This line has intercepts (0, 2) and (−8, 0). Its slope is $\frac{1}{4}$ (Figure 1-31).

1-8 EXERCISES

Sketch the graph of each of these sets.

1. $\{(x, y): 3x + 4y = 12\}$
2. $\{(x, y): 3x - 4y = 12\}$
3. $\{(x, y): 6x + y = 8\}$
4. $\{(x, y): x - 6y = 12\}$
5. $\{(x, y): x - y = 0\}$
6. $\{(x, y): x + y = 0\}$
7. $\{(x, y): 2x = 4y + 10\}$
8. $\{(x, y): y = 4\}$
9. $\{(x, y): (x - 2)(y + 2) = 0\}$
10. $\{(x, y): x = 2 \text{ or } x + y = 6\}$
11. $\{(x, y): x^2 = 9\}$
12. $\{(x, y): x^2 = y^2\}$
13. $\{(x, y): x + y = 6 \text{ and } x - y = 10\}$
14. $\{(x, y): x + y = 6 \text{ or } x + y = 10\}$
15. $\{(x, y): x + y = 6 \text{ and } x + y = 10\}$
16. $\{(x, y): (x - 1)^2 = 4\}$

1-9 HALF-PLANES

A point A on a line AB partitions the line into three sets of points:

(1) The point A.

(2) The points of the **half-line** that has endpoint A and contains B.

(3) The points of the half-line that has endpoint A and does not contain B.

The two half-lines are called **opposite half-lines** (Figure 1-32). Note that a half-line with endpoint A does not contain the point A. The

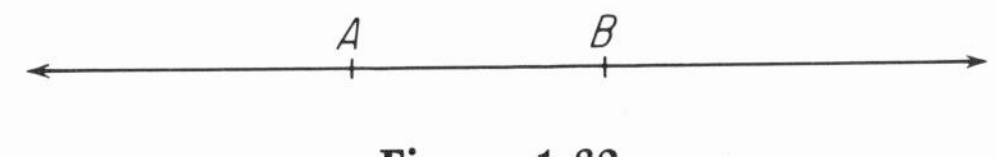

Figure 1-32

union of a half-line and its endpoint is called a **ray.** Similarly, a line on a plane partitions the plane into three sets of points: the line and two **half-planes.** The line is the **edge** of each half-plane.

Consider, for example, the line $\{(x, y): x + 2y = 6\}$. Every point of the plane is in exactly one of the three sets:

(1) The points of the line.
(2) The points of the half-plane "above" the line.
(3) The points of the half-plane "below" the line.

Each half-plane is said to be the **opposite** of the other (Figure 1-33).

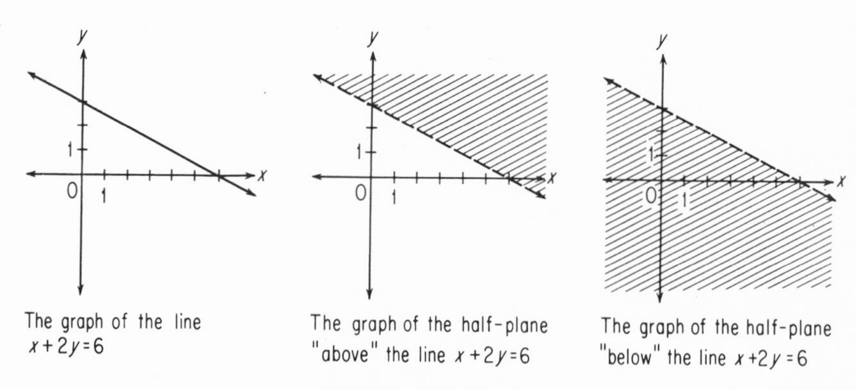

Figure 1-33

Can the points in each half-plane be described by a statement about their coordinates?

Let (a, b) denote an arbitrary point of the half-plane *below* the line $x + 2y = 6$ (Figure 1-34). Consider the line that contains the

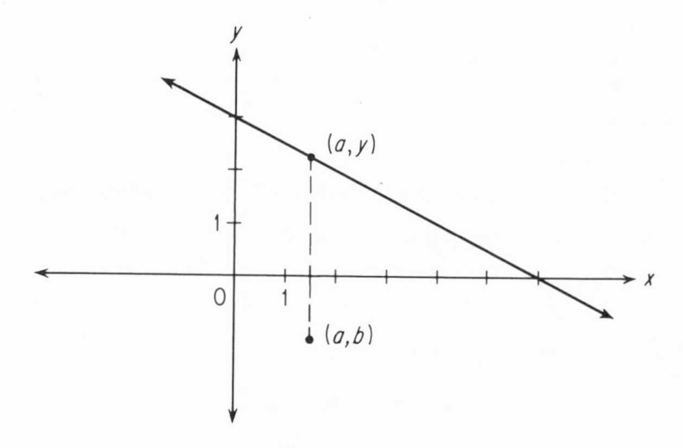

Figure 1-34

point (a, b) and is parallel to the y-axis; it intersects the line $x + 2y = 6$ at (a, y). Note that the intersection point has to have the same

x-coordinate as (a, b). The point (a, y) is on the given line, so $a + 2y = 6$. But $b < y$; therefore, $a + 2b < 6$.

A similar argument shows that if (c, d) denotes an arbitrary point of the half-plane *above* the line $x + 2y = 6$, then $c + 2d > 6$. The conclusion implied by these two observations is that the half-plane below $\{(x, y): x + 2y = 6\}$ is the graph of $\{(x, y): x + 2y < 6\}$, and the opposite half-plane is the graph of $\{(x, y): x + 2y > 6\}$.

As a second example, Figure 1-35 shows the points in the set

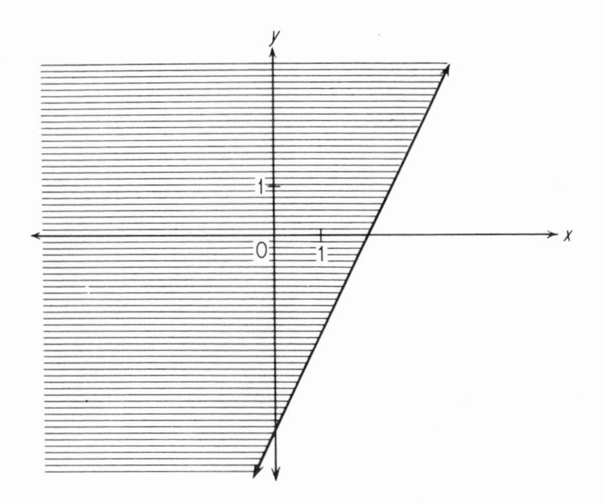

Figure 1-35

$\{(x, y): 2x - y \leqq 4\}$. Note that this set includes the points on the line $\{(x, y): 2x - y = 4\}$. The set of points is the graph of a half-plane and its edge.

Business managers sometimes use a branch of mathematics called linear programming to help them in making certain decisions. Linear programming requires the identification of points in or on the boundaries of two or more half-planes. Two examples follow.

EXAMPLE 1: Sketch the set:

$$\{(x, y): x > 0 \quad \text{and} \quad y > 0 \quad \text{and} \quad x + y < 6\}$$

A point is in this set if and only if it is in each of the half-planes $\{(x, y): x > 0\}$, $\{(x, y): y > 0\}$, and $\{(x, y): x + y < 6\}$. In Figure 1-36 the three line segments that form the boundary of this region are dashed to indicate that their points are not included.

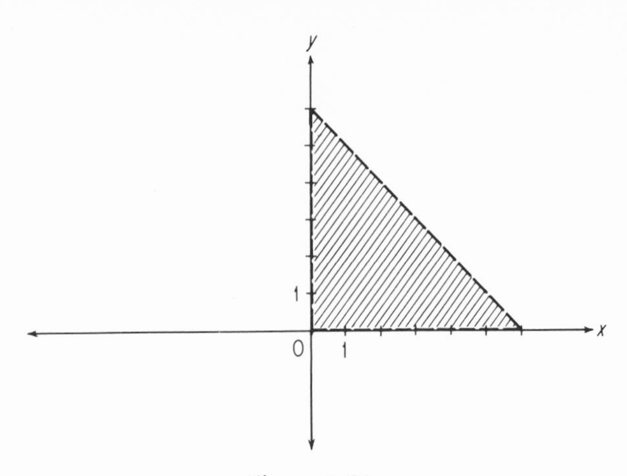

Figure 1-36

EXAMPLE 2: Sketch the set $\{(x, y) : x \geqq y \text{ and } 0 \leqq x \leqq 4\}$. The boundaries of this region (Figure 1-37) are not dashed, since boundary points are included.

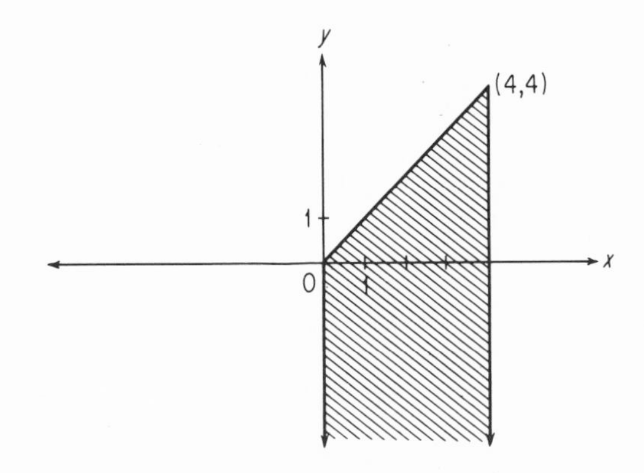

Figure 1-37

1-9 EXERCISES

Sketch each set.

1. $\{(x, y) : x > 5\}$

2. $\{(x, y) : x + y < 0\}$

3. $\{(x, y) : 2x + 3y \geqq 12\}$

4. $\{(x, y) : x - 3y \leqq 9\}$

5. $\{(x, y) : x > 0 \text{ and } y < 0 \text{ and } 2x - y \geqq 8\}$

6. $\{(x, y) : x + y \leqq 8 \text{ and } 5x - 7y \geqq 28 \text{ and } x \geqq 2\}$

1-10 FAMILIES OF LINES

There are many lines in a coordinate plane that contain the point (1, 2). All of these except the line $x = 1$ can be described by an equation of the form:

$$y - 2 = m(x - 1)$$

This equation is said to describe a **family** of lines; the arbitrary constant m is called a **parameter.** If m is chosen to be some specific number, the equation then defines a particular member of the family. Four members are shown in Figure 1-38.

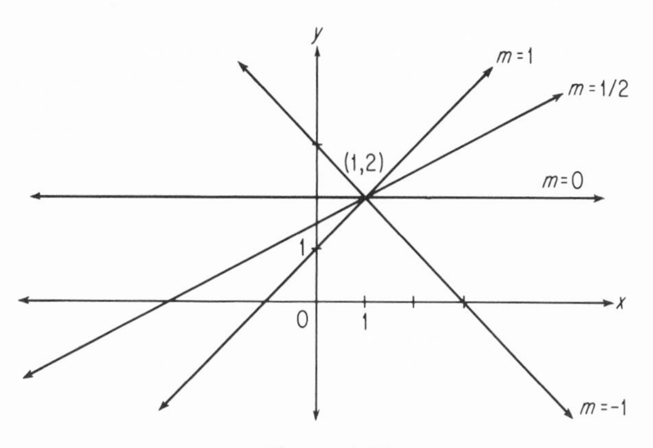

Figure 1-38

Which member of the family of lines $y - 2 = m(x - 1)$ contains the point (4, 11)? To get the answer, replace x by 4 and y by 11 in the equation $y - 2 = m(x - 1)$, since we want (4, 11) to satisfy the equation, and solve for m:

$$11 - 2 = m(4 - 1) \Leftrightarrow 9 = 3m \Leftrightarrow m = 3$$

The only solution is $m = 3$; so the required member of the family is the graph of $y - 2 = 3(x - 1)$.

The equation $y = 5x + b$, with parameter b, describes another family of lines. All members of this family have slope 5. Let b have specific values such as -3, -1, 0, and $\frac{3}{2}$ to obtain the equation of specific members of the family (Figure 1-39).

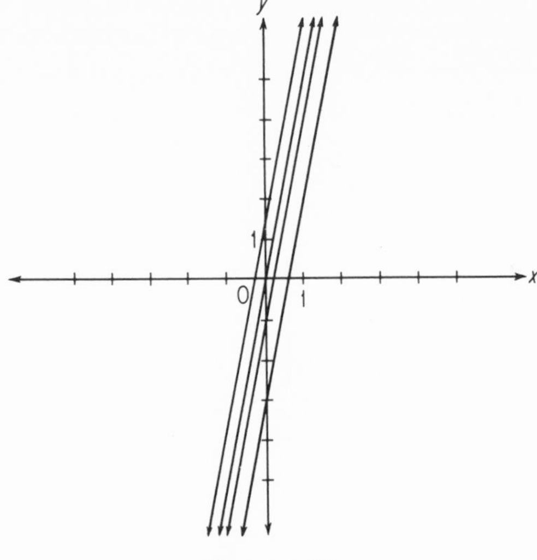

Figure 1-39

1-10 EXERCISES

1. Consider a family of lines on a coordinate plane described as follows: The family contains every line (except the line $x = 4$) that passes through the point $(4, -2)$.

(a) Write a concise set description of this family.

(b) Which member of this family also contains the origin?

(c) Which member of this family contains the point $(1, 4)$?

2. On a coordinate plane consider the lines that have slope -2.

(a) Write an equation for this family of lines.

(b) Which member of this family contains $(0, 8)$? Contains $(0, 0)$? Contains $(0, -4)$?

3. Sketch at least four different members of the family of lines $\{(x, y): x + y = k\}$. The "parameter" here is k.

(a) Which member of this family contains the origin?

(b) Which member contains $(4, -75)$?

1-11 THE DISTANCE BETWEEN TWO POINTS IN A COORDINATE PLANE

Throughout the rest of this chapter we shall specify that the coordinate planes under discussion have the same unit length on the y-axis as on the x-axis.

In Section 1-4 the distance between two points on a number line was defined. If A and B, with respective coordinates a and b, denote two points on a number line, then $d(A, B) = |a - b|$. The reader will recall that for arbitrary points A, B, and C on a number line:

(1) $d(A, B) \geqq 0$.
(2) $d(A, B) = 0 \Leftrightarrow A = B$.
(3) $d(A, B) = d(B, A)$.
(4) $d(A, B) \leqq d(A, C) + d(C, B)$.

If a coordinate plane has the same unit length on the y-axis as on the x-axis and if P and Q denote any two points in the plane, we would like to define "the distance between P and Q" and to make the definition consistent with the Pythagorean theorem of Euclidean geometry and consistent also with the definition of distance between points on a number line.

Suppose the coordinates of P are (a, b) and the coordinates of Q are (c, d). If $a \neq c$ and $b \neq d$, then P, Q and the point with coordinates (c, b) are the vertices of a right triangle (Figure 1-40). The

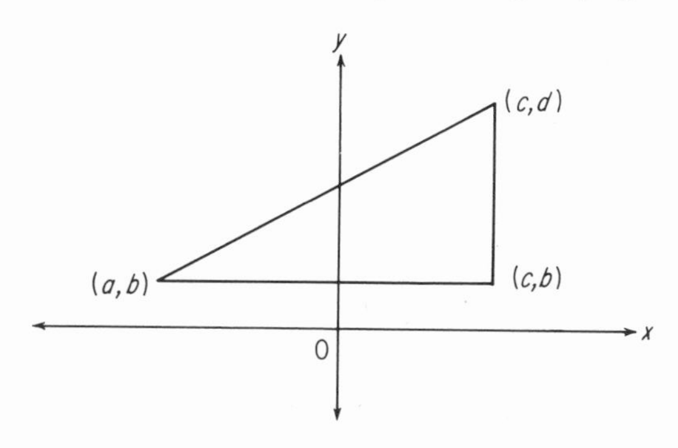

Figure 1-40

lengths of the two legs of this right triangle are $|a - c|$ and $|b - d|$. Therefore, by the theorem of Pythagoras, the length of the hypotenuse is $\sqrt{(a - c)^2 + (b - d)^2}$.

The immediately preceding result suggests a definition for distance between two points in a coordinate plane.

Definition: Suppose the coordinates of P are (a, b) and the coordinates of Q are (c, d). Let $d(P, Q)$ denote the distance between P and Q. Then

$$d(P, Q) = \sqrt{(a - c)^2 + (b - d)^2}.$$

For example, if P has coordinates $(7, -7)$ and Q has coordinates $(3, -4)$, then

$$\begin{aligned} d(P, Q) &= \sqrt{(7-3)^2 + [-7-(-4)]^2} \\ &= \sqrt{4^2 + (-3)^2} \\ &= 5 \end{aligned}$$

Observe that in the definition of the distance between the two points $P(a, b)$ and $Q(c, d)$ it is not required that $P \neq Q$, or that $a \neq c$, or $b \neq d$. The reader should satisfy himself that if P and Q are points on a line that is parallel to one of the coordinate axes, then $d(P, Q)$, as now defined, is consistent with the definition of the distance between points on a number line. It is also true (and the reader is asked to prove this in the exercises) that if P, Q, and R are arbitrary points in a coordinate plane, then

(1) $d(P, Q) \geqq 0$.
(2) $d(P, Q) = 0 \Leftrightarrow P = Q$.
(3) $d(P, Q) = d(Q, P)$.
(4) $d(P, Q) \leqq d(P, R) + d(R, Q)$.

1-11 EXERCISES

1. Find the distance between the points with coordinates:
(a) $(0, 4)$ and $(0, -3)$
(b) $(5, 1)$ and $(9, 1)$
(c) $(6, 0)$ and $(0, 6)$
(d) $(7, -1)$ and $(4, 3)$
(e) $(1, 6)$ and $(-1, -1)$
(f) $(21, 1)$ and $(-4, 101)$

2. In a coordinate plane, what is the distance between
(a) The origin and an arbitrary point (x, y)?
(b) The point with coordinates (h, k) and an arbitrary point (x, y)?

3. Find the length of the line segment that has endpoints:
(a) $(3, 1)$ and $(11, 7)$
(b) $(\sqrt{63}, 1)$ and $(0, 2)$

4. Show that the triangle with vertices at $(3, 0)$, $(6, 10)$, and $(9, 0)$ is an isosceles triangle.

5. Use the converse of the Pythagorean theorem to show that the triangle with vertices at $(5, 6)$, $(13, 6)$, and $(5, 12)$ is a right triangle.

6. Prove that if P and Q denote arbitrary points in a coordinate plane, then

(a) $d(P, Q) \geqq 0$.
(b) $d(P, Q) = 0 \Leftrightarrow P = Q$.
(c) $d(P, Q) = d(Q, P)$.

(d) $d(P, Q) \leqq d(P, R) + d(R, Q)$. (You may use an informal geometric argument here.)

7. An object moves along a path in a coordinate plane in such a way that the coordinates (x, y) of its position at time t are given by

$$x(t) = 1 - 2t \qquad y(t) = 3 - t$$

(a) What are the coordinates of the object's position when $t = 0$? $t = 5$?
(b) Write the rule that expresses the distance from the object to the origin at time t.
(c) Write the rule that expresses the distance from the object to the point with coordinates $(2, 3)$ at time t.
(d) Sketch the path of this object from the time $t = 0$ to the time $t = 4$.

8. (a) A line segment has endpoints with coordinates $(3, 6)$ and $(3, 10)$. Find the midpoint of this line segment.
(b) A line segment has endpoints with coordinates $(1, 4)$ and $(5, 8)$. Find the midpoint of this line segment.
(c) A line segment has endpoints (a, b) and (c, d). Show that the coordinates of the midpoint of the segment are $((a + c)/2, (b + d)/2)$.

1-12 CIRCLES AND DISKS

Consider the graph of $\{(x, y): x^2 + y^2 = 25\}$. The equation

$$x^2 + y^2 = 25$$

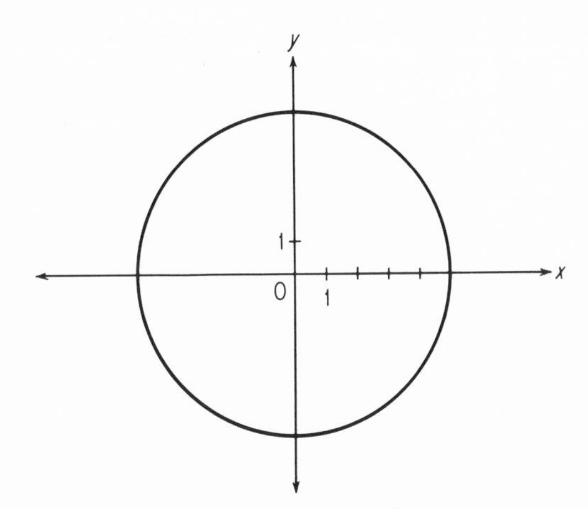

Figure 1-41 $\{(x, y): x^2 + y^2 = 25\}$

is equivalent to the equation

$$\sqrt{x^2 + y^2} = 5.$$

Therefore, the graph of $\{(x, y): x^2 + y^2 = 25\}$ consists of those points (and only those points) that are 5 units from the origin. The graph is a circle with center at the origin and radius 5 (Figure 1-41).

In general, if h and k denote any real numbers and r denotes an arbitrary positive number, then the graph of

$$\{(x, y): (x - h)^2 + (y - k)^2 = r^2\}$$

is a circle with center at $(h; k)$ and radius r. For example, the graph of

$$\{(x, y): (x - 2)^2 + (y + 1)^2 = 4\}$$

is a circle with center at $(2, -1)$ and radius 2 (Figure 1-42). The equation

$$(x - 2)^2 + (y + 1)^2 = 4$$

is called an **equation of the circle.**

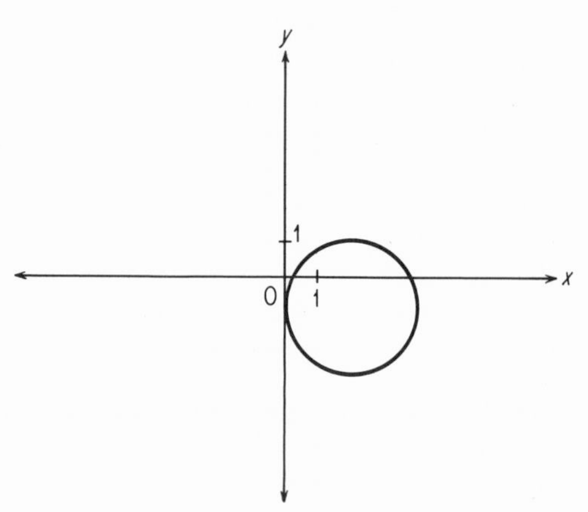

Figure 1-42 $\{(x, y): (x - 2)^2 + (y + 1)^2 = 4\}$

All the points that are *inside* a given circle (but not *on* the circle) constitute a set called an **open disk.** For example, the graph of the set $\{(x, y): (x - 1)^2 + y^2 < 9\}$ is an open disk (Figure 1-43). The center of the disk is $(1, 0)$ and the radius is 3.

A **closed disk** consists of all the points *inside* or *on* a given circle. The graph of $\{(x, y): (x - 1)^2 + y^2 \leqq 9\}$ is a closed disk (Figure 1-43).

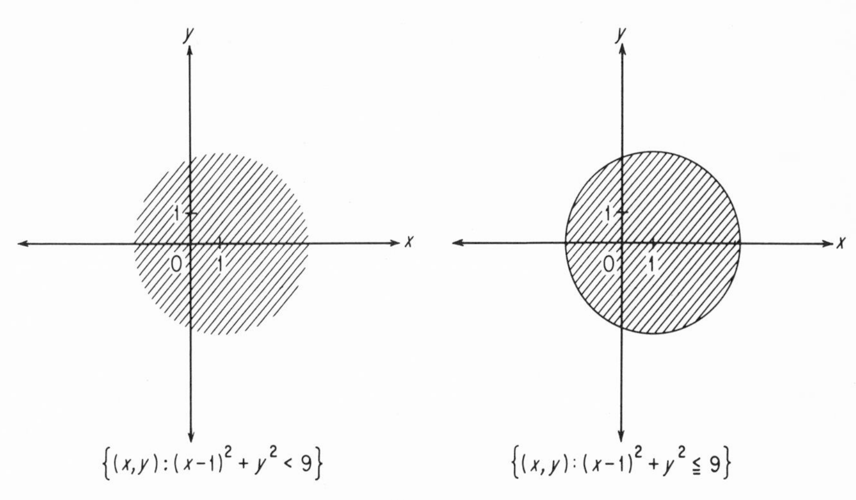

Figure 1-43

1-12 EXERCISES

1. A circle with radius 1 is called a **unit** circle. Write an equation of the unit circle whose center is

(a) The origin of a coordinate plane.
(b) The point (3, 5).

2. Write an equation of the circle that

(a) Has radius 2 and center (3, 2).
(b) Has radius 4 and center $(-3, 5)$.
(c) Has radius 10 and center (10, 0).
(d) Has radius $\frac{1}{2}$ and center $(-5, -1)$.

3. Sketch the graph of each of these sets of ordered pairs of real numbers.

(a) $\{(x, y): x^2 + y^2 = 49\}$
(b) $\{(x, y): x^2 + y^2 < 49\}$
(c) $\{(x, y): x^2 + y^2 \leqq 49\}$
(d) $\{(x, y): (x - 2)^2 + (y + 3)^2 < 9\}$
(e) $\{(x, y): x^2 + (y + 4)^2 \geqq 25\}$
(f) $\{(x, y): x^2 + (y + 4)^2 = 0\}$
(g) $\{(x, y): 2 \leqq x^2 + y^2 \leqq 5\}$
(h) $\{(x, y): (x^2 + y^2 - 4)(y - x) = 0\}$

4. The center of an open disk is $(-2, 6)$ and the radius of the disk is 10. Give an algebraic description of the set of points of the disk.

5. In the following figure each circle has its center at the origin; the

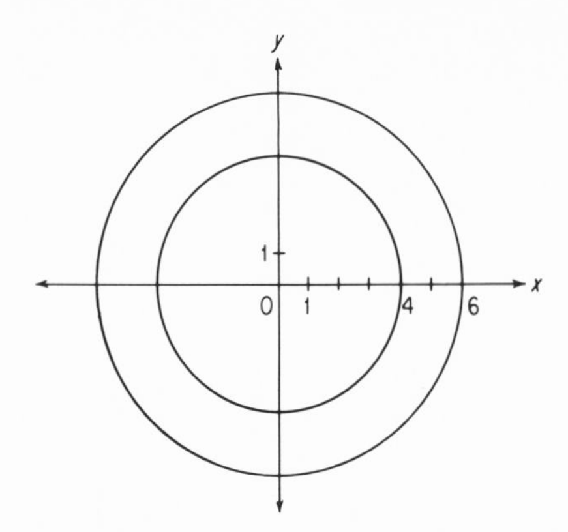

radius of the inner circle is 4 and the radius of the outer circle is 6. Write a description of a set of ordered pairs (x, y) so that the graph of the set will consist of the points that are inside the larger circle but outside the smaller circle.

2

The Idea of a Function and Simple Examples

2-1 WHAT IS A FUNCTION?

Let A and B stand for non-empty sets. Suppose a rule is given that assigns to each member of A exactly one member of B. Then the rule, together with the sets A and B, is said to be a **function from *A* into *B*.** Set A is called the **domain** of the function. All elements of B that are assigned by the rule to some member of A constitute a set called the **range** of the function.

Let A denote the set of letters $\{a, b, c\}$ and B denote the set of numbers $\{3, 5, 6, 8\}$. The diagram in Figure 2-1 defines a function

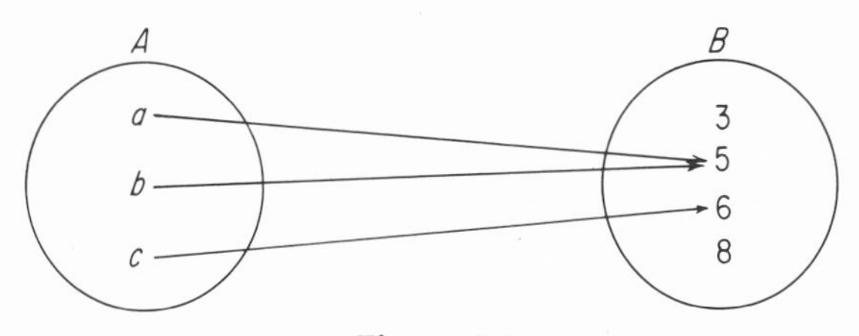

Figure 2-1

from A into B. The domain of this function is the set $\{a, b, c\}$ and the range is $\{5, 6\}$.

A function may be considered as a **mapping** of the elements of its domain onto the elements of its range. In the previous example we say, "a is mapped onto 5, b is mapped onto 5, and c is mapped onto 6." We also say that under the function "the **image** of a is 5, the image of b is 5," etc.

EXAMPLE 1: Suppose with every real number x we associate the number $2x + 1$. This defines a function from the set of real numbers into the set of real numbers. The function maps 2 onto 5; it maps 0 onto 1, etc.

EXAMPLE 2: Suppose the teacher of a certain class expects to assign one of the letters A, B, C, D, F to each member of the class. Then he plans to construct a function with the domain the set of students in the class. The range will be a subset (perhaps all) of the set $\{A, B, C, D, F\}$.

The definition of a function is so broad that it allows for many different kinds of functions. In fact, their number is uncountable.

The domain and range are frequently sets of *numbers*. But the definition of a function does not require this. The domain (or range) of a particular function may be a set of people, or geometric points, or ordered pairs of numbers, or any other nonempty set of elements.

The rule for a function may have a variety of forms: It may be an algebraic formula, but it may instead be a table of values, a diagram, or a word description.

It is important to remember that a particular function is defined by supplying answers to two questions:

(1) What is the domain of the function?

(2) How does one obtain the image of an arbitrary element of the domain?

Whatever its form, the rule for determining the image of an element under a function must never assign two or more different images to the same element in the domain. For example, the diagram in Figure 2-2 does *not* define a function from A into B because it maps c onto two different elements in B.

A rule that associates two or more different images with the same element in the domain might be an important rule, but we just do not call it the rule for a *function*. (Such rules are said to define relations. See Section 2-6.)

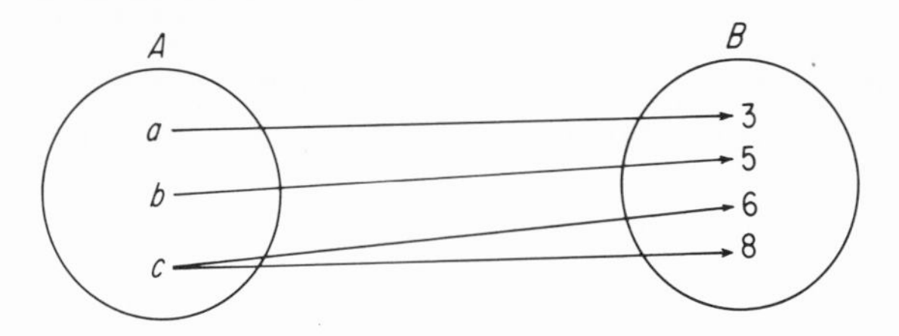

Figure 2-2

It is customary to denote functions by single letters such as f, g, and h. If x is any member of the domain of function f, then $f(x)$ is the name for the element assigned to x by the function f.

EXAMPLE 3: Let the function f have for its domain the set of real numbers and let f assign to x the number $3x + 5$. Then

$$\begin{aligned} f(1) &= 3(1) + 5 = 8 \\ f(-1) &= 3(-1) + 5 = 2 \\ f(a) &= 3(a) + 5 = 3a + 5 \\ f(-a) &= 3(-a) + 5 = -3a + 5 \\ f(b + 2) &= 3(b + 2) + 5 \\ f(x) &= 3x + 5. \end{aligned}$$

Is there an element in the domain of f that has 38 as its image?

$$\begin{aligned} f(x) = 38 &\Leftrightarrow 3x + 5 = 38 \\ &\Leftrightarrow 3x = 33 \\ &\Leftrightarrow x = 11 \end{aligned}$$

Therefore the number 11 has 38 as its image.

Two functions f and g are said to be **equal** if and only if they have the same domain A and for each x belonging to A, $f(x) = g(x)$.

EXAMPLE 4: Let the functions f and g be defined as follows:

$$f(x) = x(x - 1), \qquad \text{domain: the real numbers}$$

$$g(x) = x^2 - x, \qquad \text{domain: the real numbers}$$

Then f and g are equal functions since

$$x(x - 1) = x^2 - x$$

for all real numbers x.

EXAMPLE 5: Suppose functions f and g are defined by:

$f(x) = 3x + 1$, domain: the set of integers

$g(x) = 3x + 1$, domain: the set of real numbers

Then f and g are *different* functions, even though they have the same rule.

A function f has been compared to a machine (Figure 2-3) that has an input and an output. An element x of the domain is fed into

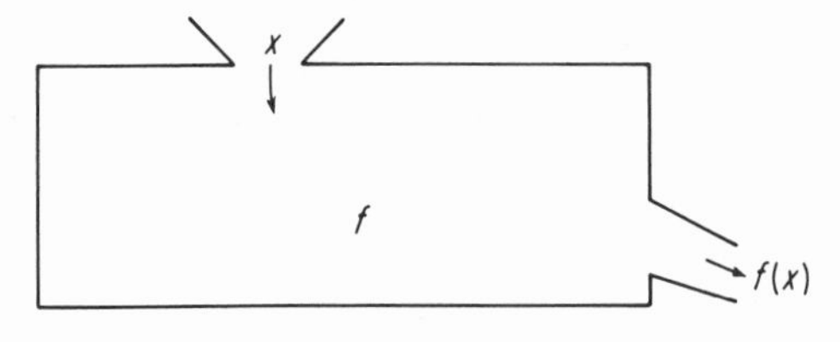

Figure 2-3

the function machine; the machine "works on it," and out comes $f(x)$. The set of elements that are fed into a particular machine is called the **domain** of the function; the set of all elements that are produced is called the **range** of the function.

2-1 EXERCISES

1. Given that f is the function whose domain is the set of positive integers and which pairs with each positive integer n the integer $5n$.

(a) What is the image of 3 under the function f?
(b) $f(4) = ?$
(c) $f(10) = ?$
(d) What element in the domain of f has 35 as its image?
(e) For what n is $f(n) = 10$?
(f) Is there some element in the domain that maps onto 100?
(g) Is there some element in the domain that maps onto 33?
(h) What is the range of f?

2. Suppose f has the set of all real numbers as its domain and has the rule $f(x) = x^2 + 1$.

(a) $f(3) = ?$
(b) $f(-3) = ?$
(c) Is 0 in the range of f?
(d) Is 26 in the range of f? If so, find all elements that map onto 26.

3. You may have made extensive use of the division algorithm of arithmetic: If an integer n is divided by a positive integer d, then there exist unique integers q and r such that $n = qd + r$, where $0 \leqq r < d$. Now suppose g is the function that has as its domain the set of all integers and has as its rule that $g(n)$ is the unique remainder obtained when n is divided by 5.

(a) $g(12) = ?$
(b) $g(-17) = ?$
(c) What is the range of g?
(d) List at least three different integers that map onto 4.

4. Suppose a plane is flying along a straight line at a uniform speed of 200 miles per hour. Point A is on the flight path, and at time $t = 0$ the plane is 100 miles distant from point A and flying toward A. Let s stand for the distance in miles between the plane and point A and let t stand for time measured in hours.

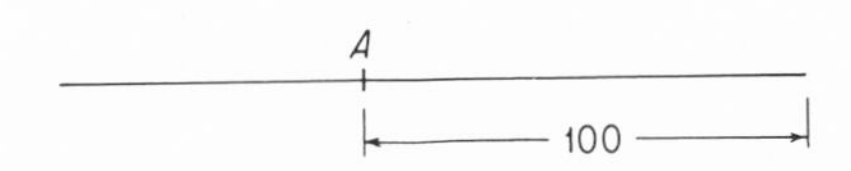

(a) Write the rule for the function that associates with an arbitrary time $t \geqq 0$ the distance s of the plane from point A.
(b) $s(4) = ?$
(c) When will the plane be 500 miles from A?

5. A salesman has this agreement about his salary: Each month he is to receive \$200 plus 5% of the dollar value S of the sales he makes that month.

(a) Let P stand for the amount of his monthly paycheck in dollars and and write the rule for a function that shows how P depends upon S.
(b) $P(S)$ will be \$1000 provided what is true about S?
(c) $P(S)$ will be *at least* \$1000 provided what is true about S?

6. Let S denote the function that has as its domain the set of all positive integers and has as its rule that $S(n)$ is the sum of the first n positive integers.

(a) $S(3) = ?$
(b) $S(5) = ?$
(c) Do you know or can you devise a way of stating the rule for this function so as to simplify the calculation of $S(n)$ for large values of n?
(d) $S(1000) = ?$

7. Let P be a function that has as its domain the set of all positive integers and has as its rule that $P(n)$ is the number of primes less than or equal to n.

(a) $P(1) = ?$
(b) $P(6) = ?$
(c) If $P(n) > 10$, then what conclusion can be drawn about n?

8. Each of these questions should suggest a function with which you are familiar from your previous studies. In each case give the appropriate rule for the function.

(a) How do you find the area of a disk if you know the radius?

(b) How do you find the surface area of a cube if you know the length of each edge?

(c) How do you find the circumference of a circle if you know the radius?

(d) How do you find the radius of a circle if you know the circumference?

(e) If an object moves at a uniform speed of s ft/sec along a straight line, how far does it travel in t seconds?

(f) If an object falls from rest in a vacuum, how far does it fall in t seconds?

(g) If an object falls from rest in a vacuum, what is its speed after t seconds?

9. Suppose the set of real numbers is the domain for each of two functions f and g. Suppose $f(x) = |x|$ and $g(x) = \sqrt{x^2}$. Does $f = g$? Explain.

10. The rule for a function may be given by a table of values. The following function has been used by married taxpayers filing joint returns. We call it a tax function F.

Tax Rate Schedule

If the amount of your taxable income I is		*The amount of your tax F(I) is*
not over $1,000		14% of I
over	but not over	
$1,000	$2,000	$140 + 15% of ($I$ – $1,000)
$2,000	$3,000	$290 + 16% of ($I$ – $2,000)
$3,000	$4,000	$450 + 17% of ($I$ – $3,000)
$4,000	$8,000	$620 + 19% of ($I$ – $4,000)
$8,000	$12,000	$1,380 + 22% of ($I$ – $8,000)
$12,000	$16,000	$2,250 + 25% of ($I$ – $12,000)
...	...	...

(a) $F(1200) = ?$
(b) $F(3500) = ?$
(c) $F(8250) = ?$
(d) $F(825) = ?$

2-2 THE GRAPH OF A FUNCTION

For many important functions both the domain and the range are sets of numbers. For such functions the number pairs $(x, f(x))$ can be plotted as points in a plane (a rectangular coordinate system being used). We shall call the resulting geometric figure **the graph of the function.**

EXAMPLE 1: Suppose the domain of f is $\{-1, 1, 3\}$ and the rule is that each number n is mapped onto $2n$. Then the graph of f consists of the three points indicated in Figure 2-4.

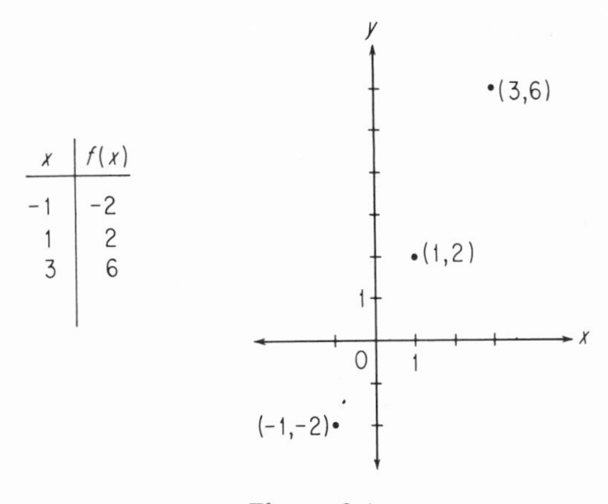

x	$f(x)$
-1	-2
1	2
3	6

Figure 2-4

Sometimes the domain of a function is an interval of real numbers. In sketching the graph of such a function it is frequently justified to plot a few well-chosen points and then to assume that the graph is a smooth curve passing through these points.

EXAMPLE 2: Draw the graph of the function defined by $f(x) = x^2 - 2$ for $0 \leqq x \leqq 3$. To do this choose several specific values of x and compute $f(x)$. We know that Figure 2-5 shows part of the graph. It is not easy to prove in every detail that the complete graph is a smooth curve that is steadily rising from left to right. But intuition strongly suggests that such is the case. On the basis of this assumption (unproved here, but correct) we

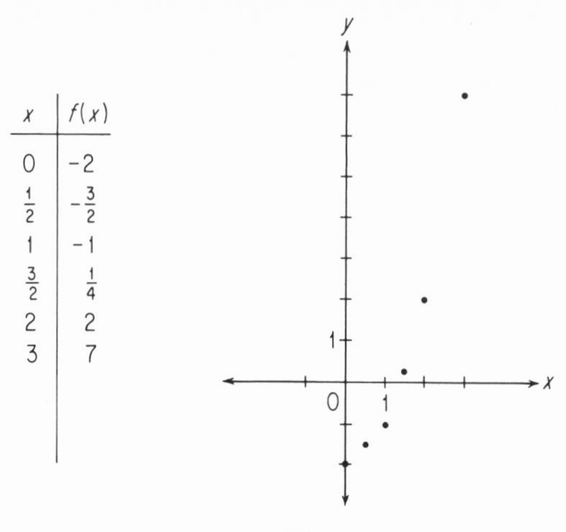

x	$f(x)$
0	-2
$\frac{1}{2}$	$-\frac{3}{2}$
1	-1
$\frac{3}{2}$	$\frac{1}{4}$
2	2
3	7

Figure 2-5

fill in the rest of the graph. Observe that the range of the function can be shown as a set of points along the y-axis; the domain can be shown as a set of points along the x-axis (Figure 2-6).

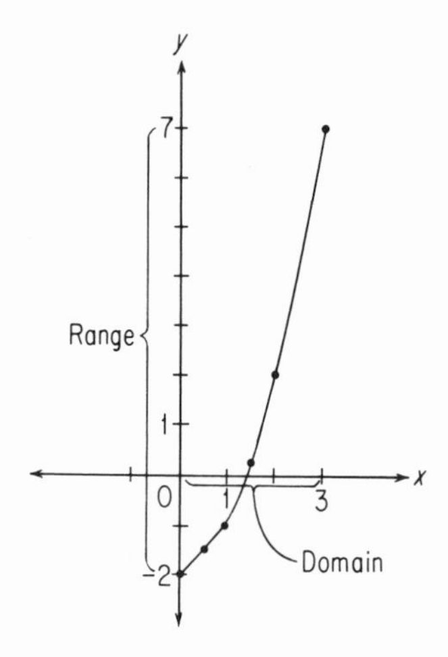

Figure 2-6

An important feature of the graph of every function is this: If the domain is represented along the x-axis, then every line that contains a point of the domain on the x-axis and is parallel to the y-axis will intersect the graph of f in exactly one point. This result follows from the definition of a function, which requires that each x in the domain have a *unique* image.

A circle (Figure 2-7) cannot be the graph of a function, since elements of the domain that are between -1 and 1 have two images, a positive image and a negative image. But a semicircle (Figure 2-8) can be the graph of a function.

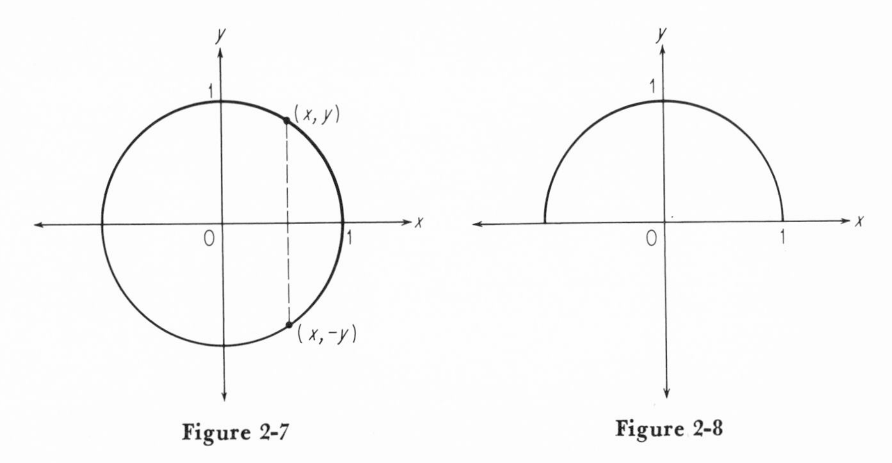

Figure 2-7 **Figure 2-8**

Sometimes a graph is used to define a function. Imagine a device that automatically and continuously records the temperature at a certain weather station. The information produced over a 24-hour period might appear as in Figure 2-9.

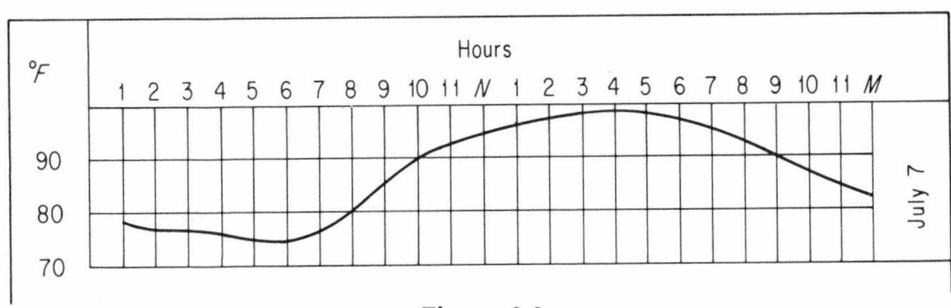

Figure 2-9

Notice that the graph in Figure 2-9 is the graph of a function, the domain is shown, and the graph serves as the rule of the function. The domain of the function is the set of numbers that identify the time (along the horizontal axis). The rule may be used as follows: To determine the temperature at a specified time t (in the domain of the function) draw a vertical line v that contains the point with coordinate t on the horizontal (time) axis. This line will intersect the temperature graph at a unique point. Through this point draw a horizontal line h. The coordinate of the point at which the line h intersects the vertical axis denotes the recorded temperature at the specified time t.

In general, let G denote a set of points of a coordinate plane. Suppose G has the property that every line parallel to the y-axis intersects G in at most one point. Then G is the graph of a function. In Figure 2-10 the set of points G is the graph of a function. The domain is $\{x: -1 \leqq x \leqq 3\}$ and the range is $\{y: 1 \leqq y \leqq 5\}$.

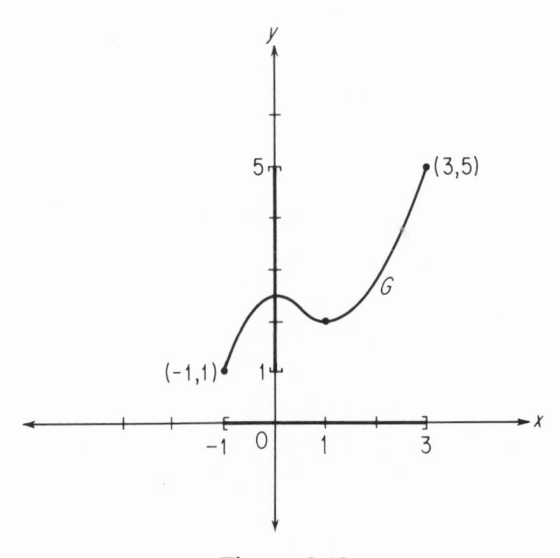

Figure 2-10

2-2 EXERCISES

1. You may assume that the graph of each of the functions in this exercise is a smooth curve with no "gaps." Plot enough discrete (individual) points for each function so that the pattern of the graph is clear, and then complete the graph by drawing a smooth curve through these discrete points. Describe the range of each of these functions.

(a) $f(x) = 2$; domain: all real numbers.
(Note: This function is called a **constant function.**)
(b) $f(x) = 4 + 3x$; domain: $\{x: -1 \leqq x \leqq 5\}$.
(c) $f(x) = x^2 + 1$; domain: $\{x: x \geqq 0\}$.
(d) $f(x) = -x$; domain: all real numbers.
(e) $f(x) = 25 - x^2$; domain: $\{x: -5 \leqq x \leqq 5\}$.
(f) $f(x) = \dfrac{1}{x}$ for $x > 0$.
(g) $C(F) = \frac{5}{9}(F - 32)$ for $32 \leqq F \leqq 212$.

(h) $f(x) = 3$ for all x.
(i) $f(x) = x^3$ for all x.

2. The adjoining graph is a straight line segment with endpoints $(-2, 3)$ and $(1, 4)$. This graph defines a function f that has $\{x: -2 \leqq x \leqq 1\}$ as its domain.

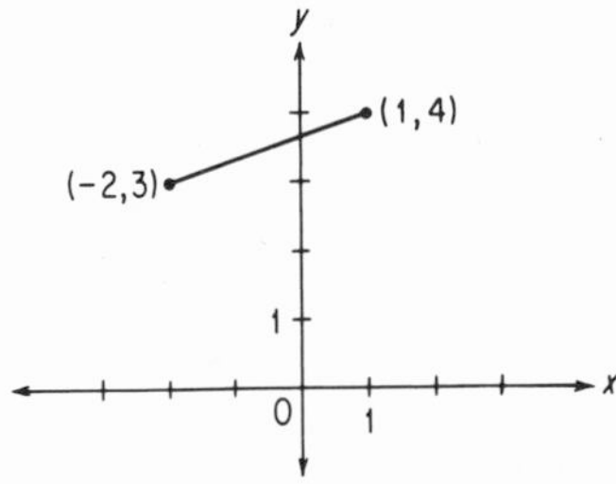

(a) What is the range of f?
(b) $f(-2) = ?$
(c) Since you are told that this graph is a line segment, develop an algebraic rule for f.

3. For some of the graphs presented in this exercise it is true that the indicated set of points is the graph of a function (with the domain of the function represented as a set of points on the x-axis). Select those graphs for which this is true. Explain why each of the others is *not* the graph of a function that has its domain represented as a set of points on the x-axis.

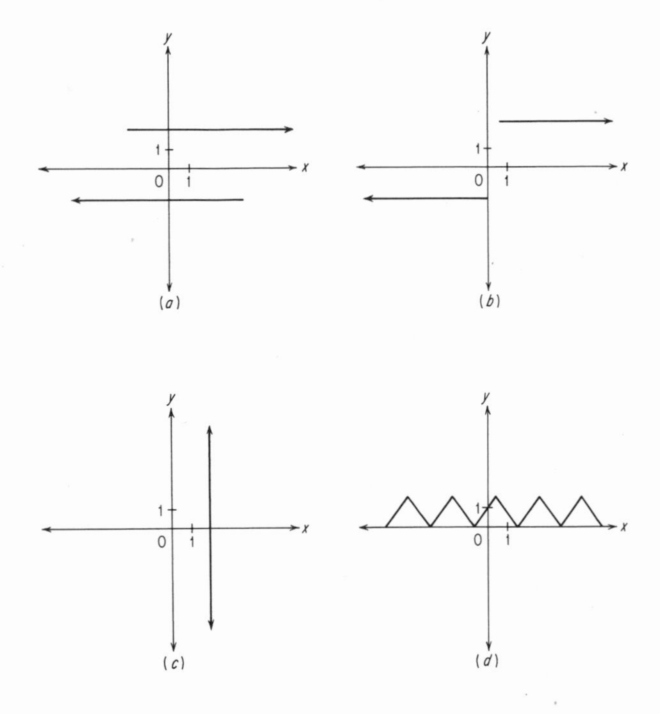

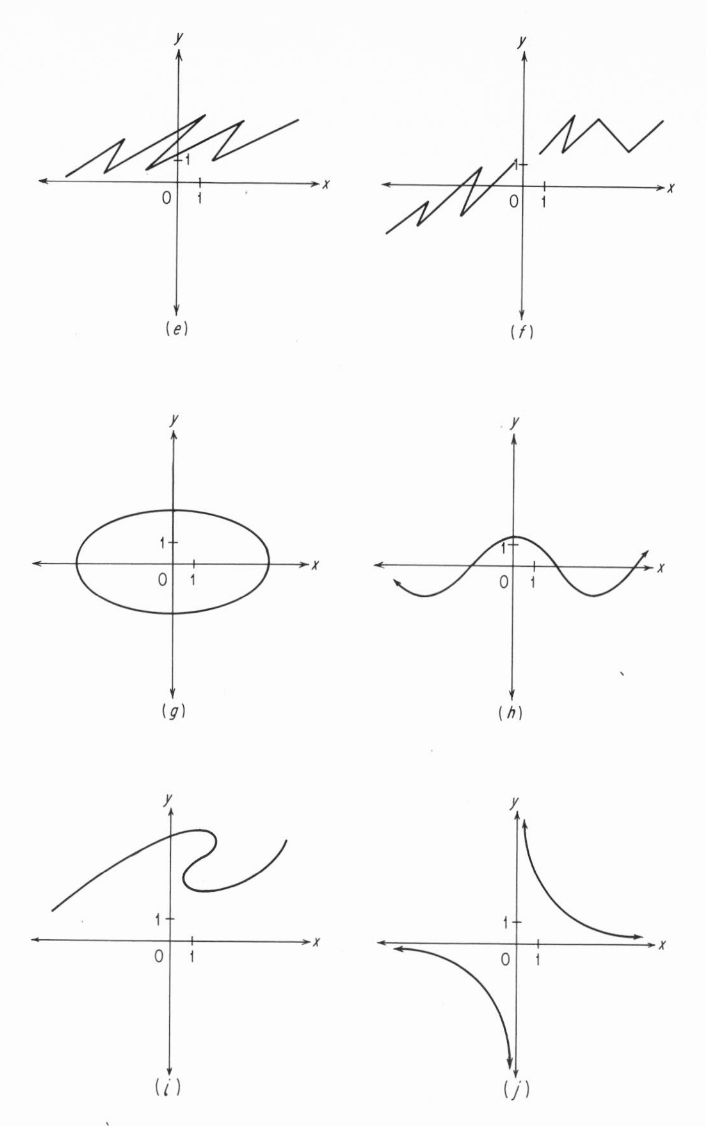

2-3 MORE ON THE WAY A FUNCTION IS SPECIFIED

Suppose a function has the set of all real numbers as its domain and has the rule $f(x) = 2x + 1$. If we change this rule to read $y = 2x + 1$, or $s = 2t + 1$, or $z = 2w + 1$, the notation has been changed but the rule for the function is the same as ever. This function still maps 3 onto 7, it maps 0 onto 1, etc.

The student must become accustomed to such variations in the way a function is written. As long as the domain is clear and the rule provides a unique image for each element in the domain, then the function is a **well-defined function.**

In addition to such variations in notation, the student must become accustomed to a sometimes puzzling but long established convention in defining functions: If just the rule is given with no statement about the domain, you should make one or both of these assumptions:

(1) The author wishes you to judge from the context of the discussion what domain is appropriate.

(2) The author means for the domain to be the most inclusive set of real numbers for which the rule "makes sense."

EXAMPLE 1: In a discussion about spheres, an author refers to "the function $V(r) = (\frac{4}{3})\pi r^3$," with no mention of the domain and no implied boundaries on r except that it stands for the radius of a sphere. You would assume that the domain of the function is the set of all positive numbers.

EXAMPLE 2: An author refers to the function

$$y = \frac{1}{x - 1}$$

No mention is made of the domain, or of an application of the function. You would assume that the domain is the set of all real numbers except 1. The rule will not produce an image for 1, but it will produce an image for other real numbers.

This practice of omitting mention of the domain can cause trouble. Two functions with the same rule but different domains may have quite different properties. Nevertheless, the practice is widespread and the student must learn to live with it. Usually, this convention leads to no trouble. But it is something to be careful about.

2-3 EXERCISES

Each of these equations may be interpreted as the rule for a function. However, in each case the specification of a particular function is incomplete because the domain is not mentioned. Assume that the writer intends each domain to be the most inclusive set of real numbers to which the rule can apply (and always gives a real number as the image). On this basis, describe each domain.

1. $f(x) = \dfrac{3}{x+2}$

2. $y = \sqrt{x}$

3. $y = \dfrac{x+1}{x-2}$

4. $t = \dfrac{1}{s}$

5. $g(x) = \dfrac{1}{x^2 - 4}$

6. $h(x) = \sqrt{x+5}$

7. $t = \dfrac{s}{s^2+4}$

8. $t = \sqrt{s^2}$

2-4 THE ZEROS OF A FUNCTION

Let f be a function from a subset of the real numbers into the real numbers. If $f(c) = 0$, where c is a number in the domain of f, then c is called a **zero of f.** Thus a zero of f is the same as a solution (or root) of the equation $f(x) = 0$.

EXAMPLE 1: Let f be defined by $f(x) = x - 2$; domain: the real numbers. Then 2 is the only zero of f.

$$\begin{aligned} f(x) = 0 &\Leftrightarrow x - 2 = 0 \\ &\Leftrightarrow x = 2 \end{aligned}$$

EXAMPLE 2: Let g be defined by $g(x) = x(x - 2)$; domain: the real numbers. Then the zeros of g are 0 and 2.

$$\begin{aligned} g(x) = 0 &\Leftrightarrow x(x-2) = 0 \\ &\Leftrightarrow x = 0 \quad \text{or} \quad x = 2 \end{aligned}$$

This statement means that $g(0) = 0$ and $g(2) = 0$, but no other real number is mapped onto 0 by g.

Since a zero of a function is an element in the domain whose image is 0, we can associate it with the point $(c, 0)$ on the graph of the function. Each such point is on the x-axis and is called an x-intercept of the graph.

Some functions have many zeros; others have exactly one; some have no zeros at all. Sometimes it is easy to locate the zeros; sometimes it is difficult. In working some of the exercises that follow you will need to recall these three basic facts about numbers:

(1) $x^2 \geqq 0$ for *every* real number.

(2) $x \cdot y = 0 \Leftrightarrow x = 0$ or $y = 0$; that is, the product of two numbers is 0 if and only if one of the factors is 0.

(3) Any expression $ax^2 + bx + c$, where a, b, and c are given real numbers and x represents a real number, can be written in the form $a(x - r)(x - s)$ in at most one way except for the order of the factors; that is, *factorization is unique except for the order of the factors.* For $r \leqq s$ and $p \leqq q$, $(x - r)(x - s) = (x - p)(x - q) \Leftrightarrow r = p$ and $s = q$.

2-4 EXERCISES

1. The domain of each function in this exercise is the set of all real numbers. Find all the zeros of each function.

(a) $f(x) = 6x + 8$ (b) $g(x) = \dfrac{2}{x^2 + 1}$

(c) $y = x^2 + 5$ (d) $y = x^2 - 5$

(e) $y = (x - 7)(x - 4)$ (f) $y = x(x - 3)$

(g) $y = x^3 + 1$ (h) $y = 2x^2 + 3$

(i) $y = (x + 1)^2(x - 1)$ (j) $y = \dfrac{x + 4}{x^2 + 1}$

2. Here are the graphs of some functions. Name the zeros (if any) of each function.

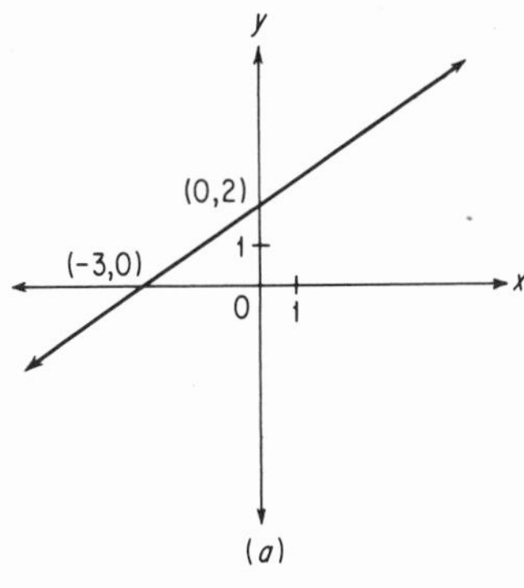

(a)

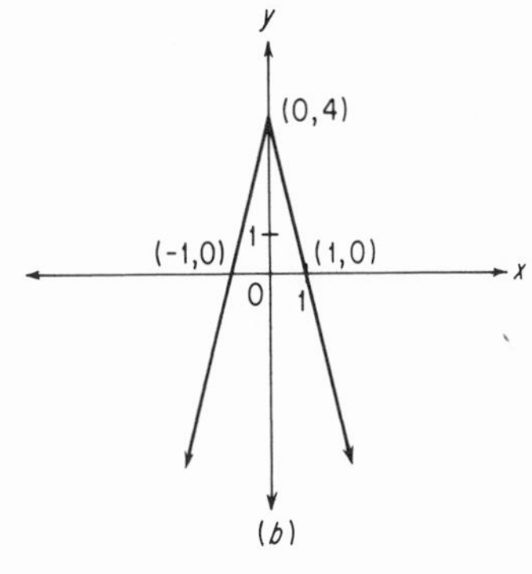

(b)

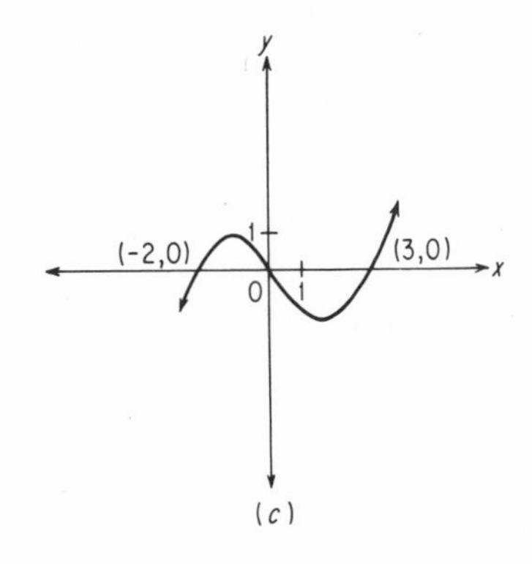

(c)

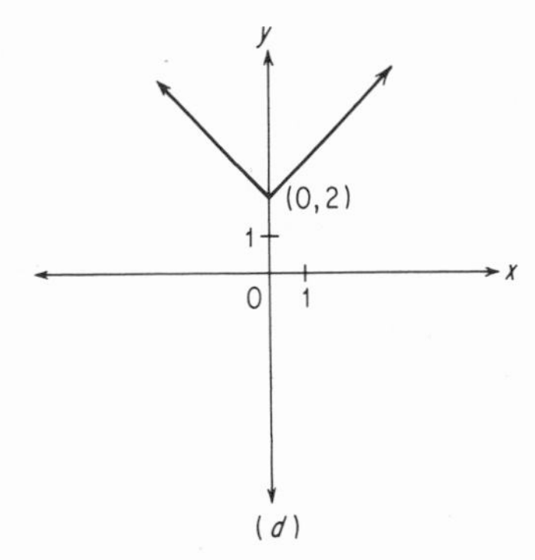

(d)

2-5 FUNCTIONS DEFINED BY A MULTI-PART RULE

For some functions the rule is composed of several parts (corresponding to different subsets of the domain). Such a rule is called a multi-part rule. The tax function of Section 2-1, Exercise 10, is an important example. Other examples follow.

EXAMPLE 1: Suppose f has the set of all real numbers as its domain, and

$$f(x) = \begin{cases} x, & \text{if } x \text{ is negative} \\ 2, & \text{if } 0 \leqq x \leqq 1 \\ 2x, & \text{if } x > 1 \end{cases}$$

This rule yields an image for every real number. The graph of f is drawn in Figure 2-11. The origin $(0, 0)$ is not a point of this

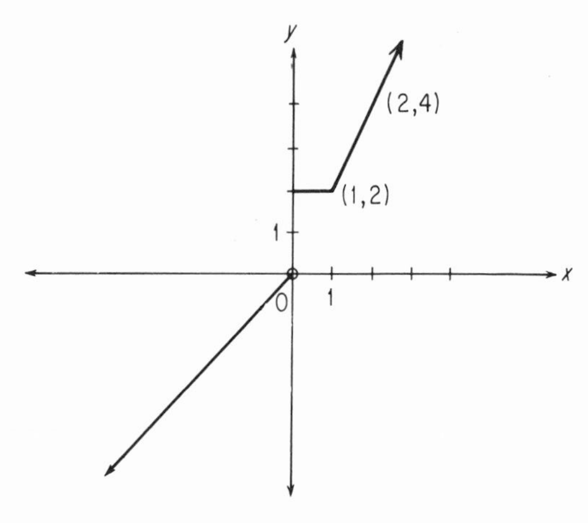

Figure 2-11

graph, so a small circle has been placed around the origin to indicate that it is not included.

EXAMPLE 2: A salesman has the following agreement about his monthly pay: He receives \$200 plus 15% of the dollar value *in excess of \$1000* of his sales for the month. We are interested in describing the rule for the function that relates the dollar value of his monthly pay to the dollar value of his monthly sales. Let P

denote his monthly pay in dollars and S denote his monthly sales in dollars. Then

$$P(S) = \begin{cases} 200, & \text{if } 0 \leqq S \leqq 1000 \\ 200 + 0.15(S - 1000), & \text{if } S > 1000 \end{cases}$$

2-5 EXERCISES

1. With reference to Example 2 of this section, what does the salesman earn in a month in which his sales total \$980? \$4000?

2. Sketch the graph of each of these functions.

(a) $f(x) = \begin{cases} -1, & \text{if } x \text{ is negative} \\ 1, & \text{if } x \text{ is not negative} \end{cases}$

(b) $g(x) = \begin{cases} -2, & \text{if } x < 1 \\ x, & \text{if } x \geqq 1 \end{cases}$

(c) $h(x) = \begin{cases} x, & \text{if } x \text{ is positive or zero} \\ -x, & \text{if } x \text{ is negative} \end{cases}$

(d) $F(x) = \begin{cases} x^2, & \text{if } x \text{ is negative} \\ -x^2, & \text{if } x \text{ is not negative} \end{cases}$

3. A familiar function relates the weight of a letter to the cost of sending it by first class mail to any place in the United States or Canada. The rule is a multi-part rule, usually presented in table form.

Weight (*ounces*)	*Total Cost* (*cents*)
1 or less	6
2 or less but more than 1	12
3 or less but more than 2	18
4 or less but more than 3	24
etc.	

No parcels over 20 pounds (320 ounces) can be accepted for first class mail.

(a) What is the domain of this function?

(b) Sketch the graph with the domain restricted so that the weight is not more than 5 ounces.

4. Let f be a function that has the set of all real numbers as its domain and has the rule $f(x) = [x]$ (read "bracket x"), where $[x]$ is defined to be the greatest integer less than or equal to x. Then

$$f(\tfrac{1}{2}) = [\tfrac{1}{2}] = 0$$

$$f(-1) = [-1] = -1$$

$$f(3.9) = [3.9] = 3, \text{ etc.}$$

(a) $f(1.1) = ?$ $f(\pi) = ?$ $f(-2.7) = ?$
(b) What is the range of f?
(c) Sketch the graph of f.
(d) Notice the similarity of the graph of this function to the graph of the "postal" function of Exercise 3. Can you use the bracket notation to give the rule for the postal function in compact form? If not, how close can you come?

2-6 THE CARTESIAN PRODUCT OF TWO SETS; RELATIONS

Let A and B stand for non-empty sets and suppose a rule is given that assigns to each member of A exactly one member of B. Then the rule, together with the sets A and B, is called a function from A into B.

The preceding paragraph should sound familiar to you. It is the basic definition of *function* given on the first page of this chapter. However, there is a second approach to the definition of *function* that sheds additional light on the meaning of the word. To make this definition we need first to define the *Cartesian product* of two sets.

Let A and B denote sets. The collection of all ordered pairs (a, b) such that a is an element of A and b is an element of B is called the **Cartesian product** of A and B and is denoted by $A \times B$. Every subset of $A \times B$ is called a **relation on $A \times B$.**

EXAMPLE 1: Let $A = \{a, b, c\}$ and $B = \{1, 2\}$. Then $A \times B$ is the set

$$\{(a, 1), (a, 2), (b, 1), (b, 2), (c, 1), (c, 2)\}.$$

Any subset of $A \times B$, such as $\{(a, 2), (b, 1), (b, 2)\}$, is a relation on $A \times B$.

EXAMPLE 2: Let $B = \{1, 2\}$. Then $B \times B$ is the set

$$\{(1, 1), (1, 2), (2, 1), (2, 2)\}$$

EXAMPLE 3: Let R denote the set of real numbers. Then $R \times R$ consists of all ordered pairs (x, y) of real numbers. This is the set of ordered pairs that suggested to mathematicians the more general idea of Cartesian product that has been defined here. The set $R \times R$ has the entire coordinate plane as its graph and is the Cartesian product that is used most frequently in this book.

Every subset of $R \times R$ is a relation on $R \times R$ and has a graph on a coordinate plane. For example, the relation $\{(x, y): x < y\}$ has a half-plane as its graph (Figure 2-12).

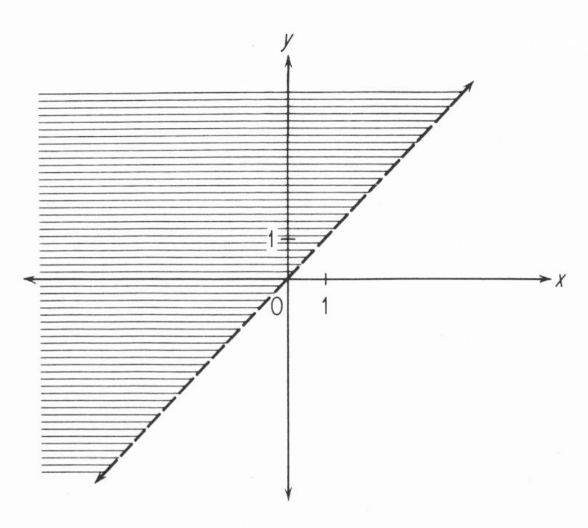

Figure 2-12

2-6 EXERCISES

1. Let $A = \{1, 2, 3\}$ and $B = \{5, 6\}$. List all the elements in the Cartesian product $A \times B$.

2. The set of all first elements of a relation on $A \times B$ is called the **domain** of the relation. The set of all the second elements is called the **range.** Let A and B denote the sets of Exercise 1.

(a) What is the domain of the relation $\{(1, 5), (1, 6), (2, 5)\}$? What is the range?

(b) Use the convention that we described for functions in Section 2-3 and find the domain of the relation on $A \times B$ that is described as $\{(a, b): b = a + 3\}$. What is the range?

3. (a) Suppose A is a set that contains exactly 2 different elements and B is a set that contains exactly 4 different elements. How many different ordered pairs are in $A \times B$?

(b) Suppose A is a set that contains exactly p different elements and B is a set that contains exactly q different elements. How many different ordered pairs are in $A \times B$?

4. Let R denote the set of real numbers. Sketch the graph of the following relations on $R \times R$. Describe the domain and the range of each.

(a) $\{(x, y): y = x\}$.
(b) $\{(x, y): y < x\}$.
(c) $\{(x, y): y = 3\}$.
(d) $\{(x, y): x + y = 4\}$.
(e) $\{(x, y): x + y \geqq 4\}$.

2-7 FUNCTIONS FROM A INTO B

We are now ready to make a second definition of *function*—a definition of a function as a special case of a relation.

Suppose A and B denote non-empty sets. Let f denote a subset of $A \times B$; that is, f is a relation on $A \times B$. Then ***f* is a function from *A* into *B*** if and only if each a belonging to A occurs as the first element of exactly one ordered pair (a, b) in f. The **domain of *f*** is A; **the range of *f*** is the subset (possibly all) of B that is the set of all the second elements of the ordered pairs of f. The phrase "function from A into B" specifies that the domain of the function is A; it also specifies that the range is a subset of B, but not necessarily all of B.

EXAMPLE 1: Let $A = \{1, 2, 3\}$ and $B = \{7, 8, 9\}$. Consider three subsets of $A \times B$:

$$f = \{(1, 7), (2, 7), (3, 8)\}$$
$$g = \{(1, 7), (1, 8), (2, 7), (3, 9)\}$$
$$h = \{(2, 7), (3, 8)\}$$

Of these three subsets of $A \times B$ only f is a function from A into B. The range of f is $\{7, 8\}$. The relation g is *not* a function, because it contains both of the pairs $(1, 7)$ and $(1, 8)$. The relation h is not a function *from A into B*, because it does not have A as its domain. It is, however, a function from A^* into B where $A^* = \{2, 3\}$.

EXAMPLE 2: Let $A = \{0, 1, 2, 3, 4, 5, 6, 7, 8, 9\}$. Consider three subsets of $A \times A$:

$$f = \{(x, y): y = x\}$$
$$g = \{(x, y): y = 6\}$$
$$h = \{(x, y): x = 6\}$$

Then f and g are both functions from A into A; f consists of ten ordered pairs and so does g. (The function g is called a **constant** function.) However, the third relation h is not a function, because it does not meet the requirement that each element of A occurs as the first element of exactly one ordered pair of h.

EXAMPLE 3: Let R denote the real numbers. Consider four subsets of $R \times R$:

$$f = \{(x, y): x + y = 6\}$$
$$g = \{(x, y): y = 6\}$$
$$h = \{(x, y): x \text{ is an integer and } y = 2x\}$$
$$F = \{(x, y): x^2 + y^2 = 25\}$$

The sets f and g are both functions from R into R; the graph of each is a line. The set h is not a function *from R into R* because its domain is just the set of integers, but it is a function *from the set of integers into R*. Finally, the set F is a relation, but not a function—it contains both (3, 4) and (3, −4) and many other different pairs that have the same first element.

Chapter 2 ends as it began—with a definition of "function." The reader can verify for himself that anything called a *function from A into B* by our new definition would also be called a *function from A into B* by our first definition, and conversely. The "rule" of the first definition is replaced by the "ordered pairs" of the second. It sometimes helps to think of a function as a specified set of ordered pairs.

2-7 EXERCISES

1. Let $A = \{1, 2, 3\}$ and $B = \{a, b, c\}$.

(a) Make up two examples of functions from A into B.

(b) Make up two examples of relations on $A \times B$ that are not functions.

2. Let $A = \{0, 1, 2, 3, 4, 5, 6, 7, 8, 9\}$. Let each of the following sets of ordered pairs (x, y) be a subset of $A \times A$; that is, x denotes an element of A and so does y. Some of the following subsets are functions from A into A; some are functions from A^* into A for a suitably chosen subset A^*; some are not functions at all. Identify each as a function from A into A, a function from A^* into A for some $A^* \neq A$, or a relation that is not a function.

(a) $\{(x, y): y = x + 1\}$ (b) $\{(x, y): y = x^2\}$
(c) $\{(x, y): x = 0\}$ (d) $\{(x, y): y = 8\}$
(e) $\{(x, y): x + y = 8\}$ (f) $\{(x, y): xy = 8\}$

3. Let R denote the set of all real numbers. Sketch the graph of each of these relations on $R \times R$. Identify those that are functions (either from R into R or from some subset of R into R).

(a) $\{(x, y): y = x + 1\}$ (b) $\{(x, y): x > 0\}$
(c) $\{(x, y): y = -2\}$ (d) $\{(x, y): x = -2\}$
(e) $\{(x, y): xy = 1\}$ (f) $\{(x, y): x^2 = 9\}$
(g) $\{(x, y): y^2 = 9\}$ (h) $\{(x, y): x > 0 \text{ and } y = x\}$

3

Linear Functions

3-1 DEFINITION AND EXAMPLES

Suppose the real numbers, or some non-empty subset of the real numbers, is the domain of a function f. Suppose f has a rule that can be put in the form $f(x) = mx + b$, where $m \neq 0$. Then f is called a **linear function.** For example, each of the following equations could serve as the rule for some linear function:

$$f(x) = 5x + 9$$

$$y = -3x - 28$$

$$y = \frac{x}{5}$$

$$y = \sqrt{2}x - \frac{1}{2}$$

Linear functions are easy to work with, and there are hundreds of important applications.

EXAMPLE 1: If the centigrade temperature is known, the corresponding Fahrenheit temperature can be calculated by the rule $F(C) = \frac{9}{5}C + 32$. This is the rule for a linear function.

EXAMPLE 2: An automobile travels at the steady rate of 50 miles per hour. The distance D traveled after an elapsed time of t hours is given by $D(t) = 50t$. Again, a rule for a linear function.

EXAMPLE 3: A mail order firm prices a certain small article at 10¢ apiece and there is an additional charge of 50¢ for postage and handling. If a buyer orders n articles, what is the total cost to him? The answer is a rule for a linear function. $C(n) = 0.10n + 0.50$, where C is expressed in dollars.

EXAMPLE 4: Suppose a coiled spring with no weight attached to it is 5 inches long. Suppose also for a specified domain of weights it stretches $\frac{1}{4}$ inch for each pound of weight w that is attached to it. Then its stretched length L is given by the linear function rule $L = \frac{1}{4}w + 5$, where L is measured in inches and w in pounds.

3-1 EXERCISES

1. Which of the following rules define linear functions? (For each function the domain is given to be the set of real numbers.)

(a) $f(x) = \frac{1}{2}x - 7$ (b) $y = 3x^2$
(c) $y = 5$ (d) $y = x - 5$
(e) $g(x) = (x - 1)^2$ (f) $h(x) = 3 + 2(x - 5)$
(g) $y = 16 + 3x$ (h) $y = x$

2. The selling price S of an article is computed as cost C plus markup. Suppose the markup is 25% of the cost.

(a) Write a formula for S in terms of C.
(b) What is the selling price of an article that costs $60?
(c) If the selling price of an article is $100, what was its cost?

3. If $100 is invested at 4% simple interest, the accumulated value S of the original investment is a linear function of the term n in years of the investment. The rule for the function is

$$S(n) = 100(1 + 0.04n) = 100 + 4n.$$

(a) $S(2) = ?$
(b) $S(5) = ?$
(c) If $S(n) = 200$, determine n.

4. Suppose a man's tax T is 20% of the amount by which his income I exceeds $4000.

(a) Write a rule for this tax function. $T(I) = ?$

(b) What is his tax if his income is $7000?
(c) If his tax is $140, what is his income?

3-2 GRAPHS OF LINEAR FUNCTIONS

If a given linear function has the set of all real numbers as its domain, its rectangular coordinate graph is a line. To show this, recall from Section 1-7 that the graph of

$$\{(x, y): y - b = m(x - a)\}$$

is the line that contains the point (a, b) and that has slope m. Now consider the general linear function

$$f(x) = mx + b, \qquad m \neq 0, \qquad \text{domain: the real numbers.}$$

Identify $f(x)$ with y to obtain

$$y = mx + b.$$

That is,

$$y - b = m(x - 0).$$

Thus the graph of the general linear function is the graph of

$$\{(x, y): y - b = m(x - 0)\}.$$

This graph is the line that contains the point $(0, b)$ and that has slope m.

For example, what is the graph of the function that has the set of real numbers as its domain and has the rule $f(x) = 2x + 1$? It is the graph of the set

$$\{(x, y): y - 1 = 2(x - 0)\}$$

which is the line that contains the point $(0, 1)$ and has slope 2. (See Figure 3-1.)

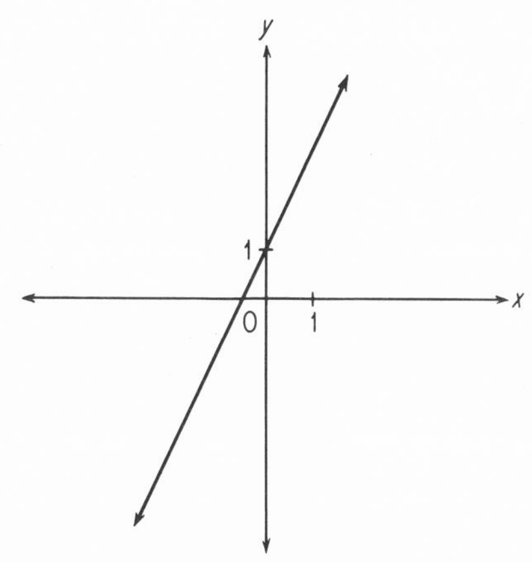

Figure 3-1

If the domain of a linear function is a proper subset of the real numbers, then its graph will be a proper subset (not all) of a line. Several examples follow.

EXAMPLE 1: Let the function g be defined by

$$g(n) = 2n + 1, \qquad \text{domain: the positive integers.}$$

This function is closely related to the function of Figure 3-1. The two rules are the same, but the domains are not. The graph of g (Figure 3-2) is a set of discrete points on the line of Figure 3-1.

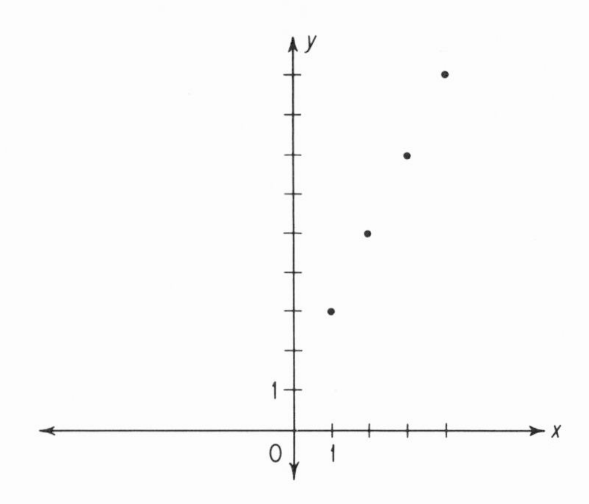

Figure 3-2

EXAMPLE 2: Let the function h be defined by

$$h(x) = 3x - 1, \qquad \text{domain: } \{x\colon x \geqq 1\}.$$

The graph of this function is a ray. It is a proper subset of the line that contains the point $(0, -1)$ and has slope 3 (Figure 3-3).

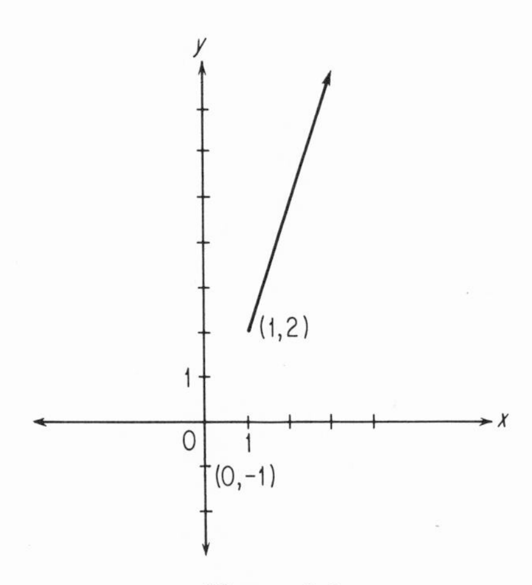

Figure 3-3

EXAMPLE 3: Let F be defined by

$$F(x) = 4 - x, \qquad \text{domain: } \{x: 1 \leqq x \leqq 3\}.$$

The graph of F is a line segment. It is a proper subset of the line that contains the point (0, 4) and has slope -1. (See Figure 3-4.)

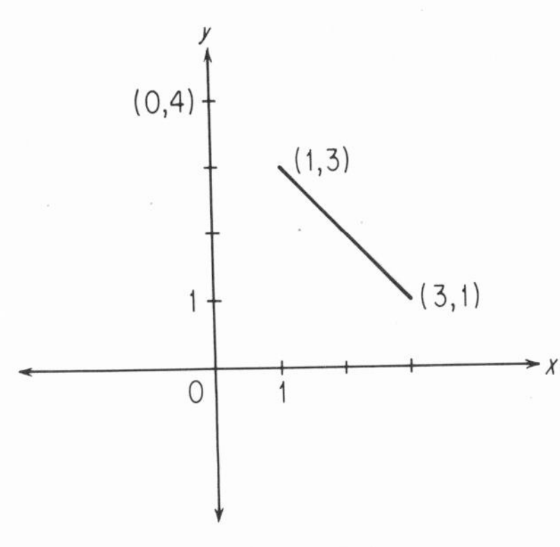

Figure 3-4

3-2 EXERCISES

1. Sketch the graph of each function.

(a) $y = 7x + 2$, domain: all real numbers
(b) $y = 4 - 3x$, domain: all real numbers
(c) $y = -(2x + 1)$, domain: all real numbers
(d) $f(n) = 5 - 2n$, domain: the positive integers
(e) $g(x) = 3x$, domain: the negative numbers
(f) $h(x) = 5x - 6$, domain: $\{x: -1 \leqq x \leqq 2\}$
(g) $y = 4 + 3x$, domain: $\{x: -2 \leqq x \leqq 0\}$

(h) $y = \begin{cases} x + 1, & \text{if } x \geqq 0 \\ x - 1, & \text{if } x < 0 \end{cases}$

(i) $y = \begin{cases} 2x, & \text{if } x > 1 \\ -x, & \text{if } x \leqq 1 \end{cases}$

2. Plot the graph of the function that relates Fahrenheit temperature to centigrade temperature C for $0 \leqq C \leqq 100$. (Show C on the horizontal axis.)

3-3 THE LINEAR FUNCTION THAT MAPS *a* ONTO *b* AND *c* ONTO *d*

How many linear functions (if any) have all these properties?

(1) The domain is all real numbers.

(2) $f(1) = 6$; that is, f maps 1 onto 6.
(3) $f(4) = 18$; that is, f maps 4 onto 18.

The answer is: There is only one such linear function. Its rule may be expressed as $f(x) = 6 + 4(x - 1)$.

The argument that supports this conclusion is brief. Since the function to be constructed is linear, its graph (Section 3-2) is the line determined by the points (1, 6) and (4, 18). By Section 1-7, the unique line that contains these points is the graph of

$$\{(x, y): y - 6 = 4(x - 1)\}.$$

Therefore, since y may be identified with $f(x)$, the required rule may be expressed as $f(x) = 6 + 4(x - 1)$.

To generalize, let (a, b) and (c, d) be arbitrary points except that $a \neq c$ and $b \neq d$. (This exception requires that the two points not be on a line parallel to either axis.) Then there is a unique linear function f that has all these properties:

(1) The domain is all real numbers.
(2) $f(a) = b$.
(3) $f(c) = d$.

The rule for this function may be expressed as $f(x) = b + m(x - a)$, where $m = (d - b)/(c - a)$. The proof of this theorem is exactly parallel to the argument for the specific example in the previous paragraph.

Another example is put in the form of a challenge. Mr. A comments to Mr. B, "I'm thinking of a linear function. All I shall tell you about it is that it maps 2 onto 8 and maps 18 onto 56. Tell me what function rule I have in mind."

Mr. B calculates briefly, then responds, "Evidently your function has the rule that x is mapped onto $8 + 3(x - 2)$."

Mr. B might also add that the rule may be expressed more simply as $f(x) = 3x + 2$.

3-3 EXERCISES

1. Suppose f is a linear function with the set of all real numbers as its domain. It maps 0 onto 8 and maps 3 onto 23.
(a) What is $f(40)$?
(b) Find the rule for f.

2. Suppose g is a linear function that maps -3 onto 0 and maps 4 onto 12. Find the rule for g.

3. Suppose h is a linear function with the set of all real numbers as its domain, $h(-1) = 4$ and $h(3) = 6$. Find the rule for h.

4. Suppose that you forget the formula for converting from centigrade to Fahrenheit, but you remember three things about it: (a) The rule that connects the two is a rule for a linear function. (b) The freezing point of water is 0°C and 32°F. (c) The boiling point of water is 100°C and 212°F. *Develop* the formula from these facts.

5. The sequence of numbers 4, 7, 10, 13, 16, . . . is said to be in arithmetic progression. This sequence suggests a function f with the set of all positive integers as its domain in which $f(1) = 4$; $f(2) = 7$; $f(3) = 10$; etc.

(a) Find $f(40)$; that is, find the fortieth term of the sequence.

(b) Write the rule for this function; i.e., write the rule for the nth term $f(n)$.

†(c) Suggest a definition of "arithmetic progression." Show that if f is a linear function with the set of all positive integers as its domain, then the set of images that f generates is an arithmetic progression.

3-4 THE SOLUTION OF LINEAR EQUATIONS IN ONE UNKNOWN

Let the domain of f be all real numbers and the rule be $f(x) = 3x + 6$. Does f map any number onto 35? To get the answer we must solve a linear equation, $3x + 6 = 35$. The solution may be written as follows:

$$3x + 6 = 35 \Leftrightarrow 3x = 29 \Leftrightarrow x = \tfrac{29}{3}$$

It is important to understand that two things are claimed when we say $\frac{29}{3}$ is *the solution* of $3x + 6 = 35$:

(1) $\frac{29}{3}$ is *a* solution; that is, $3(\frac{29}{3}) + 6 = 35$.

(2) $\frac{29}{3}$ is the *only* solution.

In Chapters 1 and 2 we have assumed that the reader knows how to solve linear equations. This assumption is probably justified for most readers—but perhaps not all. So in this section we shall review the basic properties of numbers that are useful in solving simple equations. Then we shall show by examples how these properties are applied. Anyone can become skillful in the solution of linear equations. Two things are needed: (1) an understanding of a dozen basic facts about the system of real numbers, and (2) some thoughtful practice in solving specific examples.

Here is a list of basic facts about the system of real numbers—not everything you need to know about numbers, but almost everything you need to solve linear equations. The reader should also learn the names that accompany these statements. The letters a, b, and c stand for any real numbers whatsoever (with the few important exceptions that are noted).

(1) *Closure properties.* For every a and b there is a unique sum $a + b$ and a unique product ab.

(2) *Commutative laws.*

$$a + b = b + a$$
$$ab = ba$$

(3) *Associative laws.*

$$a + (b + c) = (a + b) + c$$
$$a(bc) = (ab)c$$

(4) *Distributive laws.*

$$a(b + c) = ab + ac$$
$$(b + c)a = ba + ca$$

(5) *Additive and multiplicative identities.* There is a special number 0 with the property that $a + 0 = a$ and $0 + a = a$ for every a. No other number has this property. Also, there is a special number 1 with the property that $1 \cdot a = a$ and $a \cdot 1 = a$ for every number a. No other number has this property.

(6) *Law of the zero product.* For every a, $0 \cdot a = 0 = a \cdot 0$. Also $ab = 0$ implies $a = 0$ or $b = 0$. These two parts may be combined to make one compact statement: $ab = 0 \Leftrightarrow a = 0$ or $b = 0$.

(7) *Additive and multiplicative inverses.* Every number a has a unique additive inverse $-a$ with the property that $a + (-a) = 0$ and $(-a) + a = 0$. With the exception of 0, every number a has a unique multiplicative inverse $1/a$ with the property that $a(1/a) = 1$ and $(1/a) \cdot a = 1$. The number $1/a$ is called the **reciprocal** of a. The number 0 has no reciprocal.

(8) $-(ab) = (-a)b = a(-b)$.

(9) If $a \neq 0$, $(1/a)b = b/a$.

(10) $-(a/b) = (-a)/b = a/(-b)$.

(11) $a = b \Leftrightarrow a + c = b + c$ for every choice of a, b, and c; that is, adding the same number to each member of an equation gives an equivalent equation.

(12) If $c \neq 0$, then $a = b \Leftrightarrow ac = bc$; that is, multiplying each member of an equation by the same nonzero number gives an equivalent equation.

EXAMPLE 1: Solve the equation $4x + 7 = 12$. The steps are put down in detail and at the right are some of the justifying principles. The reader is asked to fill in the gaps in the reasoning.

$$4x + 7 = 12$$

$\Leftrightarrow (4x + 7) + (-7) = 12 + (-7)$	Add (-7) to each member
$\Leftrightarrow 4x + [7 + (-7)] = 5$	Use associative law of addition
$\Leftrightarrow 4x + 0 = 5$	
$\Leftrightarrow 4x = 5$	
$\Leftrightarrow (\frac{1}{4})(4x) = (\frac{1}{4})(5)$	Multiply each member by $\frac{1}{4}$
$\Leftrightarrow [(\frac{1}{4}) \cdot 4]x = \frac{5}{4}$	Use associative law of multiplication
$\Leftrightarrow 1 \cdot x = \frac{5}{4}$	
$\Leftrightarrow x = \frac{5}{4}$	

All these steps are involved whenever the equation $4x + 7 = 12$ is solved. However, the student should understand that ordinarily he is not expected to do this much writing. The experienced reader solves $4x + 7 = 12$ without writing anything; the less experienced may wish to write something like this:

$$4x + 7 = 12$$
$$\Leftrightarrow 4x = 5$$
$$\Leftrightarrow \quad x = \tfrac{5}{4}$$

The $\Leftrightarrow$ symbol is a valuable reminder of what is being claimed. The ultimate claims are

(1) $x = \frac{5}{4}$ implies $4x + 7 = 12$.
(2) $4x + 7 = 12$ implies $x = \frac{5}{4}$.

Each of these claims is the *converse* of the other.

EXAMPLE 2: Solve $2(x + 1) = 2x + 5$.

$$2(x + 1) = 2x + 5$$

$\Leftrightarrow 2x + 2 = 2x + 5$	Use the distributive law
$\Leftrightarrow (2x + 2) + (-2) = (2x + 5) + (-2)$	Add (-2) to each member

$\Leftrightarrow 2x + [2 + (-2)] = 2x + [5 + (-2)]$	Use the associative law of addition
$\Leftrightarrow 2x = 2x + 3$	
$\Leftrightarrow 2x + (-2x) = (2x + 3) + (-2x)$	Add $(-2x)$ to each member
$\Leftrightarrow 2x + (-2x) = (-2x) + (2x + 3)$	Use the commutative law of addition
$\Leftrightarrow 2x + (-2x) = [(-2x) + (2x)] + 3$	Use the associative law of addition
$\Leftrightarrow [2 + (-2)]x = [2 + (-2)]x + 3$	Use the distributive law
$\Leftrightarrow 0 \cdot x = 0 \cdot x + 3$	
$\Leftrightarrow 0 = 0 + 3$	Use the law of the zero product
$\Leftrightarrow 0 = 3$	

But $0 \neq 3$, so this equation has no solution.

EXAMPLE 3: Suppose $a^2x + x + a = 10$. Express x in terms of a.

$$a^2x + x + a = 10$$

$\Leftrightarrow a^2x + x = 10 - a$	Add $(-a)$ to each member
$\Leftrightarrow (a^2 + 1)x = 10 - a$	Use the distributive law
$\Leftrightarrow x = \dfrac{10 - a}{a^2 + 1}$	Multiply by the reciprocal of $a^2 + 1$

Note that the last step in this solution is justified for every real number choice of a, since $a^2 + 1$ is not 0 for *any* choice of a.

EXAMPLE 4: Solve $\pi x + 6 = 2x$.

$$\pi x + 6 = 2x$$

$\Leftrightarrow \pi x + (-2x) = -6$	Add $(-2x - 6)$ to each member
$\Leftrightarrow (\pi - 2)x = -6$	Use the distributive law
$\Leftrightarrow x = \dfrac{-6}{\pi - 2}$	Multiply by reciprocal of $\pi - 2$

3-4 EXERCISES

1. Solve each of these equations for t.

(a) $3t + 5 = 6t + 7$
(b) $2t - 1 = 5t + 19$
(c) 75% of $t = 600$
(d) $5t - 1 = 2t + 3(t + 6)$
(e) $\dfrac{t}{t+1} = 6$
(f) $\dfrac{t+1}{t} = 6$
(g) $2t = 0$
(h) $\frac{1}{2}t + \frac{3}{5}t = 8$
(i) $\sqrt{2}\,t + t = 5$
(j) $\pi t + t = 0$
(k) $\dfrac{2}{t} + \dfrac{3}{t} = 7$
(l) $\pi(t - 2) = 0$

2. Solve for x in terms of a and b. Note any restrictions on a and b that are necessary for a solution to exist.

(a) $ax + bx = 15$
(b) $ax + 15x = b$
(c) $a^2x + x = b$
(d) $abx = 10$
(e) $a + bx = 10$
(f) $b = \frac{1}{2}(x + a)$
(g) $\sqrt{2}\,x + 5 = ax$
(h) $x\sqrt{a^2 + b^2} = 40$

3. Let f be the function with domain all real numbers and rule

$$f(x) = \tfrac{1}{2}x - 4.$$

(a) For what x is $f(x) = 10$?
(b) For what x is $f(x) = 22$?
(c) What element in the domain of f has -2 as its image?
(d) What element in the domain of f is mapped onto 12?
(e) Does f have a zero? If so, what is it?

4. A familiar formula is $F = \frac{9}{5}C + 32$. Express C in terms of F.

5. Is there any temperature at which the centigrade and Fahrenheit temperatures are the same number? If so, what number?

6. Let the domain of f be all real numbers and the rule be $f(x) = 2x - 7$. Is there any x such that $f(x) = x$? If so, what number? (Such a number might be thought of as a "fixed point" of the mapping.)

7. (a) Show by example that some linear functions (with domain all real numbers) do not have a fixed point.
(b) Suppose f has the set of real numbers as its domain and has the rule $f(x) = mx + b$, where $m \neq 0$ and $b \neq 0$. What is a necessary and sufficient condition that f have a fixed point?

3-5 SOME SPECIAL PROPERTIES OF LINEAR FUNCTIONS

Suppose f is a linear function with the set of all real numbers as its domain and rule $f(x) = 2x + 3$. What is the range of f? Does it, for example, include 17?

$$f(x) = 17 \Leftrightarrow 2x + 3 = 17 \Leftrightarrow x = 7$$

So the answer is that f maps 7 onto 17; no other number is mapped onto 17.

In general, if f is a linear function defined on the domain of all real numbers and if f has the rule $f(x) = mx + b$, where $m \neq 0$, and if c is an arbitrary real number, is c in the range of f?

$$f(x) = c \Leftrightarrow mx + b = c \Leftrightarrow x = \frac{c - b}{m}$$

Therefore, $(c - b)/m$ is a real number (the only one) that is mapped onto c by the function f.

This argument produces an important property of linear functions: If the domain of a linear function is all real numbers, then the rule establishes a one-to-one correspondence from the set of real numbers onto the set of real numbers. This means that every real number is used as an image, but only once. Only one number maps onto 0; only one number maps onto 17; only one number maps onto π, etc.

In particular you should note that if f has as its domain the set of all real numbers and if its rule is $f(x) = mx + b$, $m \neq 0$, then its unique *zero* is $(-b)/m$.

Another important fact about every linear function is this: It is either increasing over its domain or it is decreasing. Figure 3-5 sug-

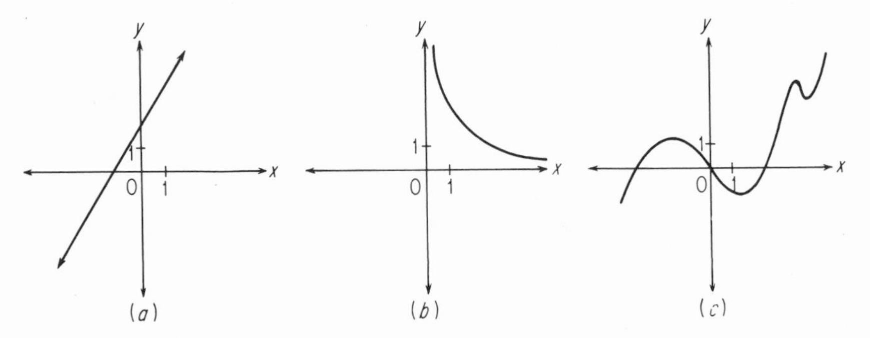

Figure 3-5

gests what is meant by these words. Part (a) shows the graph of an increasing linear function; part (b) shows the graph of a decreasing function (not linear), and part (c) shows the graph of a function (not linear) that is neither increasing nor decreasing over its domain (though it *is* increasing over some intervals and decreasing over others).

Definition: Suppose f is a function from a subset of the real numbers into the real numbers. Let c and d denote arbitrary numbers in the domain. Then f is called **increasing** if

$$c < d \quad \text{implies} \quad f(c) < f(d).$$

Also f is called **decreasing** if

$$c < d \quad \text{implies} \quad f(c) > f(d).$$

Many functions are neither increasing nor decreasing—but every linear function is one or the other, and this fact is easy to prove.

Suppose the rule for f is $f(x) = mx + b$, where m is positive (that is, the slope is positive). Let c and d stand for any two numbers in the domain that satisfy the condition $c < d$:

$$f(c) < f(d)$$

$\Leftrightarrow mc + b < md + b$

$\Leftrightarrow 0 < md - mc$ Subtracting $mc + b$ from each side

$\Leftrightarrow 0 < m(d - c)$

Since m is given to be positive and since $c < d$, it follows that

$$0 < m(d - c).$$

This is equivalent to the statement

$$f(c) < f(d).$$

Therefore, this linear function with positive slope m is increasing.

A similar argument shows that if m is negative the linear function is decreasing. Therefore, every linear function is either increasing or decreasing (since $m \neq 0$).

3-5 EXERCISES

1. Assume that the domain of each of these functions is the set of real numbers. Find the unique zero of each function.

(a) $f(x) = 5x - 7$ (b) $y = 2 + 7x$

(c) $y = 5 + 4(x - 2)$ (d) $s = \sqrt{2}t + 3$

(e) $g(x) = \frac{1}{2}x + 7$ (f) $s = 7(x - 5)$

(g) $s = 3t - \pi$ (h) $s = \sqrt{3}(x - 5)$

2. We proved that the range of a linear function is all real numbers, provided the domain is all real numbers. If the domain is restricted, so is the range. Describe the range of each of these functions.

(a) $f(x) = 3x + 1$; domain: $\{x: -1 \leqq x \leqq 7\}$.
(b) $g(x) = 2 - 5x$; domain: $\{x: 0 \leqq x \leqq 10\}$.
(c) $h(x) = 7x$; domain: $\{x: x \geqq 5\}$.
(d) $F(x) = 2x$; domain: all integers.
(e) $G(x) = 3x + 1$; domain: all integers.

3. Make up a rule for some function (with domain all real numbers):

(a) That is decreasing over the negative numbers but increasing over the positive numbers.
(b) That is neither increasing nor decreasing over any interval.

3-6 LINEAR INTERPOLATION

In Section 3-3 we proved an important theorem, which we restate here.

Suppose (a, b) and (c, d) are two points such that $a \neq c$ and $b \neq d$. Then there is a unique linear function f, with the set of all real numbers as its domain, such that $f(a) = b$ and $f(c) = d$. One form of the rule for this function is

$$f(x) = b + \frac{d - b}{c - a}(x - a).$$

If, for example, f is a linear function such that $f(2) = 6$ and $f(4) = 20$, what is the rule for f?

$$f(x) = 6 + \frac{20 - 6}{4 - 2}(x - 2)$$
$$= 6 + 7(x - 2)$$

The theorem is particularly useful in connection with a widely used procedure called **linear interpolation.** As an illustration, consider the function $f(x) = \sqrt{x}$, defined for all non-negative numbers. Suppose you know that

(1) The function is increasing.
(2) $f(600) \approx 24.495$ and $f(601) \approx 24.515$.†

† The symbol $\approx$ means "is approximately equal to."

You are asked to approximate $f(600.5)$. How would you do it?

Since f is increasing, it is reasonable to assume that even though it is not linear, it can be *approximated* by a sequence of linear functions (Figure 3-6). In particular, that part of the graph of the function on the

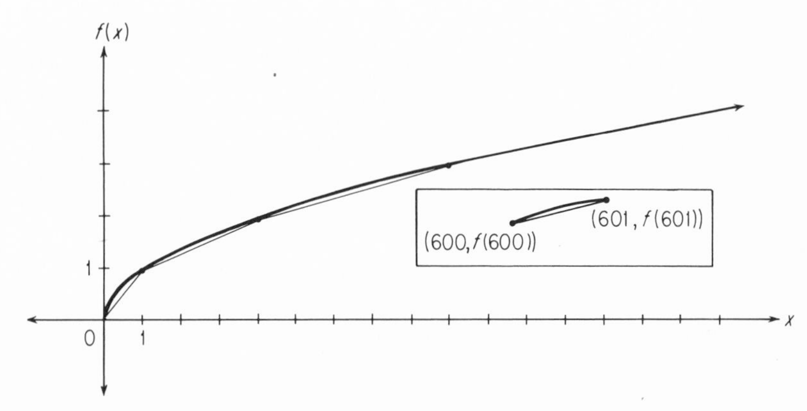

Figure 3-6 $f(x) = \sqrt{x}$

interval $\{x: 600 \leqq x \leqq 601\}$ can be approximated by a line segment that contains the points $(600, f(600))$ and $(601, f(601))$. Therefore,

$$f(600.5) \approx f(600) + \frac{f(601) - f(600)}{601 - 600}(600.5 - 600)$$

$$\approx 24.505.$$

Similarly,

$$f(600.2) \approx f(600) + \frac{f(601) - f(600)}{601 - 600}(600.2 - 600)$$

$$\approx 24.499.$$

Linear interpolation may also be used with a function that is decreasing on an interval. One such function is

$$g(x) = \cos x, \qquad \text{domain: } \left\{x: 0 \leqq x \leqq \frac{\pi}{2}\right\}.$$

Even though you may never before have heard of this function, you can quickly learn to use Table 2 (Appendix) to obtain approximate functional values of $\cos x$. We shall use Table 2 to illustrate linear interpolation.

From Table 2 we see that

$$\cos 0.26 \approx 0.9664$$

$$\cos 0.27 \approx 0.9638$$

We shall use this information and linear interpolation to approximate $\cos 0.265$ and $\cos 0.267$.

$$\begin{aligned}\cos 0.265 &\approx \cos 0.26 + \frac{\cos 0.27 - \cos 0.26}{0.27 - 0.26}(0.265 - 0.26)\\ &\approx 0.9664 - 0.0013\\ &= 0.9651\\ \cos 0.267 &\approx \cos 0.26 + \frac{\cos 0.27 - \cos 0.26}{0.27 - 0.26}(0.267 - 0.26)\\ &\approx 0.9664 - 0.0018\\ &= 0.9646\end{aligned}$$

It is not necessary that you use exactly this form for your work each time you use linear interpolation. It is important, however, that you understand what assumptions are made about a function when linear interpolation is used.

3-6 EXERCISES

1. Suppose f is a linear function with the set of all real numbers as its domain, $f(4) = 100$ and $f(12) = 120$. Use the theorem of this section to compute:

(a) $f(8)$ (b) $f(6)$

(c) $f(11)$ (d) $f(0)$

2. Suppose g is a linear function with the set of all real numbers as its domain, $g(100) = 20.34$ and $g(101) = 20.14$. Compute:

(a) $g(100.5)$ (b) $g(100.1)$

(c) $g(100.7)$ (d) $g(100.3)$

3. A widely used function called a "logarithm" function has as its domain the set of all positive numbers and a rule that is frequently written $f(x) = \log x$. Values of $f(x)$ can be approximated by the use of a table, a small portion of which is presented here.

x	$\log x$
10	1.0000
11	1.0414
12	1.0792
13	1.1139

Assuming that linear interpolation is appropriate, approximate:

(a) log 10.5 (b) log 10.1

(c) log 11.9 (d) log 12.3

4. A few entries in a table of square roots look like this.

x	$\sqrt{x}$
30	5.477
31	5.568
32	5.657
33	5.745

Use linear interpolation to approximate:

(a) $\sqrt{30.4}$ (b) $\sqrt{31.8}$

(c) $\sqrt{32.3}$ (d) $\sqrt{32.6}$

5. Use Table 2 (Appendix) and linear interpolation to approximate:

(a) cos 0.662 (b) cos 1.224

(c) sin 0.175 (d) sin 0.796

(e) tan 0.568 (f) tan 1.332

4

A Family of Absolute Value Functions

4-1 REFLECTION IN A LINE; SYMMETRY

Suppose a line l and a point P not on l are in the same plane. There is a unique point P' in the opposite half-plane from P such that l is the perpendicular bisector of the line segment PP' (Figure 4-1). The point P' is called the **reflection of *P* in *l***; P is the reflection of P' in l. If a point Q is *on* the line, we define the reflection of Q in l to be the point Q itself.

In a rectangular coordinate plane it is easy to decide what the reflection of a given point (x, y) is in the x-axis: It is the point $(x, -y)$. Similarly, the reflection of a given point (x, y) in the y-axis is the point $(-x, y)$. (See Figure 4-2.)

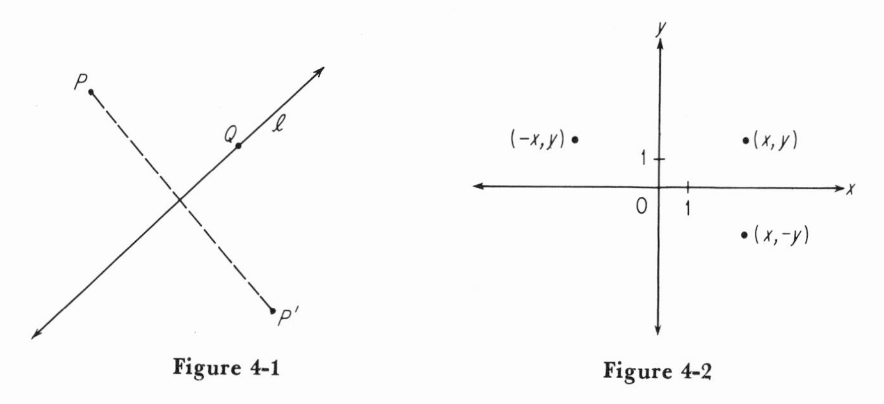

Figure 4-1 **Figure 4-2**

It is almost as easy to get the reflection of a given point (x, y) in an arbitrary line that is parallel to either of the coordinate axes. Let l denote the graph of $\{(x, y): x = h\}$, and let (a, b) denote an arbitrary point in the plane. If $a = h$, then the reflection of (a, b) in l is (a, b). But if $a \neq h$, then there is a positive number s such that $a = h + s$ or $a = h - s$, depending on whether $a > s$ or $a < s$. The reflection of the point $(h + s, b)$ in l is the point $(h - s, b)$, and conversely (Figure 4-3).

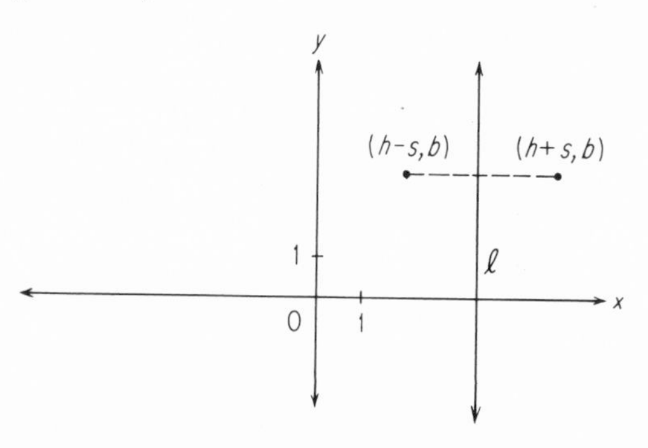

Figure 4-3

Now suppose that G denotes a set of points in a plane and l denotes a line in the same plane. Each point in G will have a reflection in l. The set G' of all these reflection points will be called the **reflection of *G* in *l*** (Figure 4-4).

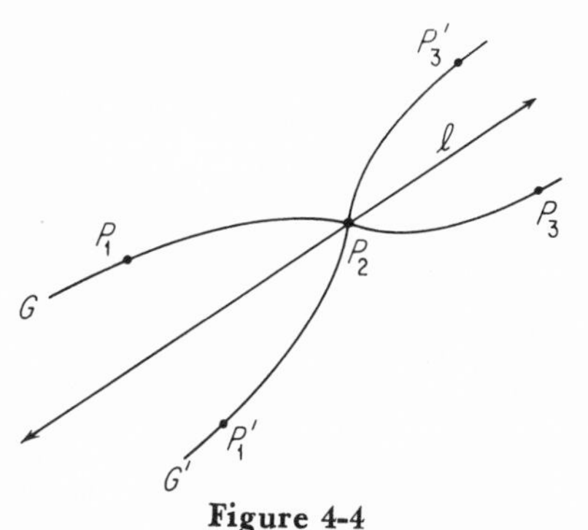

Figure 4-4

Some sets of points in a plane have the pleasing and easily understood property of being symmetric with respect to a line.

Definition: Suppose G is a set of points in a plane and l is a line in the same plane. Then G is said to be **symmetric**

with respect to l if and only if G contains the reflection of each of its points in l. We also say **l is an axis of symmetry of G.**

EXAMPLE 1: The set $\{(x, y): |x| = 2\}$ is symmetric with respect to the y-axis (Figure 4-5).

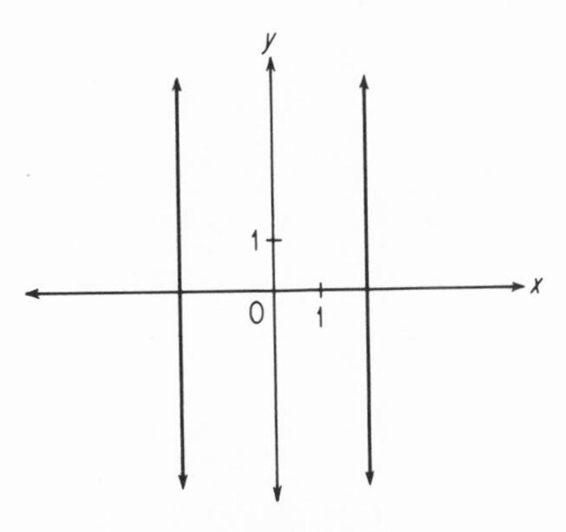

Figure 4-5

EXAMPLE 2: A circle has infinitely many axes of symmetry—each line that is in the plane and contains the center of the circle is such an axis (Figure 4-6).

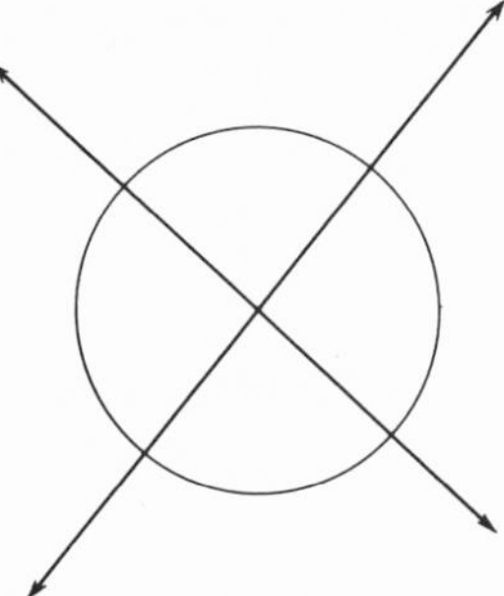

Figure 4-6

4-1 EXERCISES

1. What is the reflection of the point $(2, -5)$ in:

(a) The x-axis? (b) The y-axis?

(c) $\{(x, y): x = 3\}$? (d) $\{(x, y): y = 4\}$?

2. Let (a, b) denote an arbitrary point in the plane. What is its reflection in the line $\{(x, y): y = k\}$?

3. (a) Does a line have any axes of symmetry? Explain.
(b) Does a line segment have any axes of symmetry? Explain.

4. How many axes of symmetry does each of the following geometric figures have?
(a) A square.
(b) A rectangle that is not a square.
(c) An equilateral triangle.
(d) An isosceles triangle.
(e) A scalene triangle.
(f) An open disk.

5. Sketch the graph of the function that has the set of all real numbers as its domain and has the rule $f(x) = 2x + 3$. Then sketch its reflection in the x-axis and in the y-axis.

6. Suppose f denotes a function from a subset of the real numbers into the real numbers. Define a new function, called $-f$, as follows: (a) the domain of $-f$ is the same as the domain of f, and (b) the rule for $-f$ is $(-f)(x) = -f(x)$.
(a) Now let f be the specific function that has as its domain the set of all real numbers and has the rule $f(x) = 3x + 1$. What is the rule for $-f$? On the same set of axes sketch the graph of both f and $-f$.
(b) In general, what is the geometric relationship between the graph of f and the graph of $-f$? Explain.

7. Let l denote the line $\{(x, y): y = x\}$. In a later chapter we shall need to reflect points and sets of points in l. The reflection of point (a, b) in l is (b, a). (The student may wish to see if he can prove this.) Use this fact to sketch the reflection in l of:
(a) The line $\{(x, y): y = 2x + 3\}$.
(b) The rectangle with vertices at (4, 0), (5, 0), (4, 2), and (5, 2).

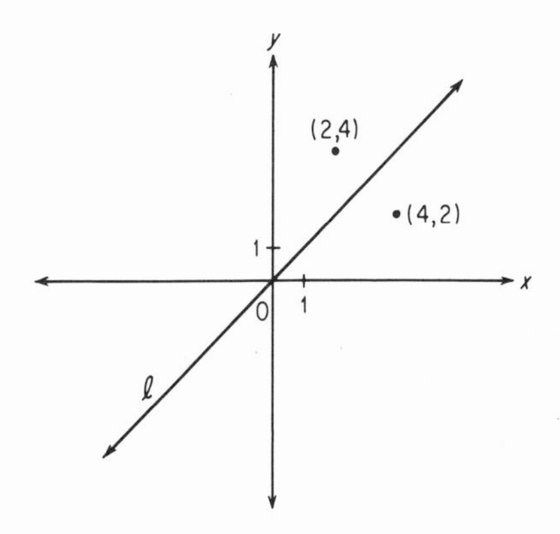

4-2 THE BASIC ABSOLUTE VALUE FUNCTION

In this chapter and the next we shall introduce many specific functions without mentioning the domain. You are to understand that the domain is the set of all real numbers or the most inclusive subset of the real numbers to which the rule of the function can be applied.

Let the rule for f be

$$f(x) = |x| = \begin{cases} x, & \text{if } x \geqq 0 \\ -x, & \text{if } x < 0. \end{cases}$$

Using the two-part form of the rule, we can easily sketch the graph of f. It is an angle—two rays with a common endpoint (Figure 4-7). The point $(0, 0)$ is called the **vertex** of the angle.

Two things about Figure 4-7 deserve special comment.

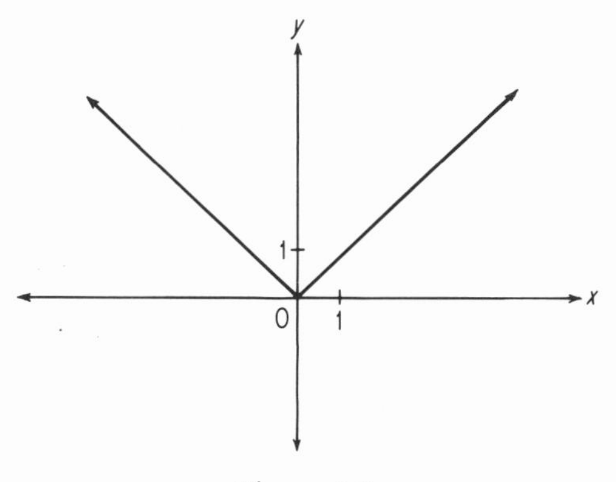

Figure 4-7

(1) The graph is symmetric with respect to the y-axis. To verify this, let $(x, |x|)$ denote an arbitrary point on the graph. Its reflection in the y-axis is $(-x, |x|)$, and this point is also on the graph since it is the same as $(-x, |-x|)$. Therefore, the graph contains the reflection of each of its points in the y-axis.

(2) The range of f is all non-negative numbers. The number 0 is in the range and it is the smallest number in that set; so 0 is called the **minimum** of the range.

The second of these two properties suggests the need of a precise definition of the minimum of a set of numbers.

Definition: Let A denote a set of real numbers. Suppose that m is a number in A and $m \leqq x$ for every x in A. Then

m is called the **minimum** of A. (Obviously, if a set has a minimum it has only one.) Similarly, if M is in A and $M \geqq x$ for every x in A, then M is called the **maximum** of A.

Some sets of numbers have neither a maximum nor a minimum; some have one but not the other; some have both. For example, the set of all real numbers has neither; the set $\{x: x \geqq 2\}$ has a minimum but not a maximum; the set $\{x: 2 \leqq x \leqq 5\}$ has both a minimum and a maximum.

The behavior of the basic absolute value function $f(x) = |x|$ can be readily understood by thinking of the distance interpretation; $|x - 0|$ gives the distance from 0 to x on a number line. This function maps each point on the line onto its distance from the origin. Now x and $-x$ are equidistant from the origin, so $f(x)$ and $f(-x)$ are the same, which accounts for the symmetry of the graph with respect to the y-axis. Also, the distance from 0 to 0 is 0. Distance between points is never negative; so 0 is the minimum number in the range of the function.

4-2 EXERCISES

1. Which of these sets has a minimum? Which has a maximum?

(a) $\{x: x < 4\}$
(b) $\{x: -2 \leqq x \leqq 0\}$
(c) $\{x: x^2 < 2\}$
(d) $\{x: 0 < x \leqq 5\}$

2. Let f have as its domain the set of all real numbers and rule $f(x) = 2 - 3x$. Does the range of f have a minimum? A maximum?

3. Let f have as its domain $\{x: 0 \leqq x \leqq 4\}$ and rule $f(x) = 2 - 3x$. Does the range of f have a minimum? A maximum?

4. Let f have as its domain the set of all positive numbers and rule $f(x) = 1/x$. Does the range of f have a minimum? A maximum?

5. On a number line the distance between two points with coordinates c and x is $|x - c|$. Use this to solve the following equations.

(a) $|x| = 6$
(b) $|x| = -1$
(c) $|x - 2| = 6$
(d) $|x + 2| = 6$
(e) $2|x - 1| = 6$
(f) $6|x - 3| = 0$
(g) $|x - 1| = |x - 5|$
(h) $|x - \pi| = |x - 10|$

6. The number 0 is the minimum number in the range of $f(x) = |x|$. What is the minimum in the range of each of the following?

(a) $g(x) = |x| + 3$ (b) $h(x) = |x| - 2$
(c) $F(x) = |x - 2|$ (d) $G(x) = |x + 3|$
(e) $H(x) = 3|x - 2|$ (f) $I(x) = |x - 2| + 5$

4-3 SOME VARIATIONS OF THE BASIC ABSOLUTE VALUE FUNCTION

In this section we shall consider four different functions, each closely related to the basic absolute value function f, but also different in an important respect. The functions to be discussed are:

$$f, \text{ defined by } f(x) = |x|$$
$$g, \text{ defined by } g(x) = |x| + 2$$
$$F, \text{ defined by } F(x) = |x - 2|$$
$$G, \text{ defined by } G(x) = 2|x|$$
$$H, \text{ defined by } H(x) = -2|x|$$

First consider g, defined by $g(x) = |x| + 2$. The graph of g, like that of f, is an angle; the two have the same axis of symmetry, but not the same vertex. Figure 4-8 shows the graphs of both functions.

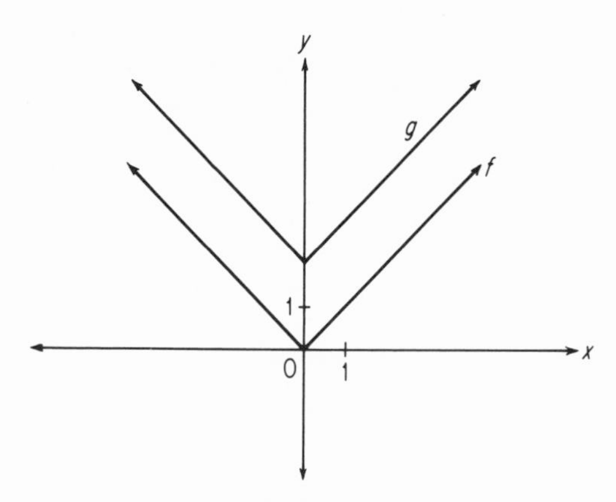

Figure 4-8

Since $g(x) = f(x) + 2$ for every x, it is as if the graph of f were made of a piece of bent wire and this were lifted up two units to form the graph of g. The correct mathematical statement is: The graph of g is congruent to the graph of f. (Two geometric figures are said to be **congruent** if one can be obtained from the other by a rigid motion, that is, a change of position that does not involve any stretching, shrinking, bending, or other "distortion" of the two figures. Unfortunately, this explanation of congruence lacks precision. How does one really check to see if two figures are congruent? A precise definition would take us too far away from the main theme of the book. We will not be able to *prove* statements about congruence, but this will not keep us from using the word where it is appropriate.)

Now consider F, defined by $F(x) = |x - 2|$. Observe that $F(x + 2) = f(x)$ for every x. This suggests that the graph of F can be obtained simply by sliding the graph of f two units to the right, keeping the vertex on the x-axis and the axis of symmetry parallel to the y-axis (Figure 4-9). This fits our intuitive concept of "rigid motion." The two graphs are congruent. The student is asked to prove (Exercise 1) that the line $\{(x, y): x = 2\}$ is an axis of symmetry of the graph of F.

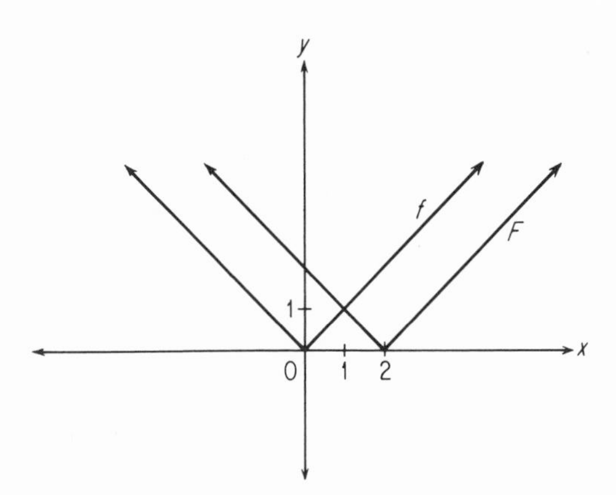

Figure 4-9

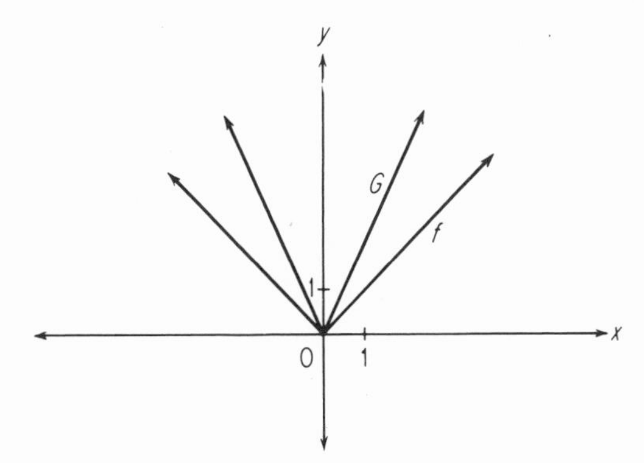

Figure 4-10

Next consider G, defined by $G(x) = 2|x|$. Observe that $G(x) = 2 \cdot f(x)$ for every x. The graphs of G and f have the same axis of symmetry and the same range, but the angles do not have the same measure; they are *not* congruent (Figure 4-10). Intuitively we feel that we can get the graph of G by grasping the rays that form the angle in the graph of f and forcing them toward each other. This changes the distance between points on the graph of f; it is not a rigid motion of the angle.

Finally, consider $H(x) = -2|x|$. Observe that $H(x) = -G(x)$ for every x. The graph of H is obtained by reflecting that of G in the x-axis. The two graphs are congruent. Also, they have the same axis of symmetry and the same vertex. However, the range of G is the non-negative numbers, in contrast to the range of H, which is all non-positive numbers. The function G has a *minimum* number in its range (and no maximum), whereas H has a *maximum* number in its range (and no minimum)(Figure 4-11).

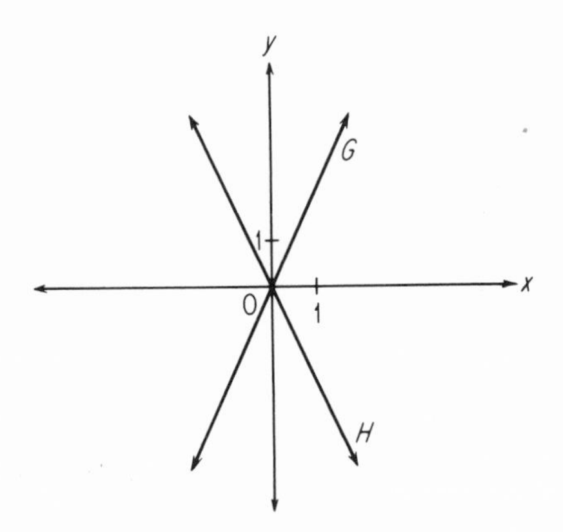

Figure 4-11

4-3 EXERCISES

1. Prove that the line $\{(x, y): x = 2\}$ is an axis of symmetry of the graph of F defined by $F(x) = |x - 2|$. [Hint: Compare $F(2 + s)$ with $F(2 - s)$ for an arbitrary real number s.]

2. The graph of each function described below is an angle that is symmetric to a line of the form $\{(x, y): x = h\}$. Sketch these graphs. For each, label the axis of symmetry and the vertex.

(a) $y = |x| - 2$ (b) $y = |x + 2|$
(c) $y = 3|x + 2|$ (d) $y = -|x + 2|$
(e) $y = |x + 2| + 1$ (f) $y = 4 + |x|$

3. It is a fact (although we have not proved it) that if a line and a ray are in the same plane and are not perpendicular, then the reflection of the ray in the line is another ray. Use this fact in answering the questions that follow.

(a) Let f be defined by $f(x) = |x|$. Draw the reflection of the graph of f in: (1) the x-axis, (2) the line $\{(x, y): x = 4\}$, (3) the line $\{(x, y): y = -2\}$.

(b) Each graph drawn in part (a) is the graph of a function. Write a rule for each function.

4. (a) Let F be defined by $F(x) = |x - 2|$. Draw the reflection of the graph of F in: (1) the x-axis, (2) the line $\{(x, y): x = -1\}$, (3) the line $\{(x, y): y = 4\}$.

(b) Each graph drawn in part (a) is the graph of a function. Write a rule for each function.

4-4 A THREE-PARAMETER FAMILY OF FUNCTIONS

The functions we have been discussing belong to a "family" described by the rule $f(x) = a|x - h| + k$. The symbols a, h, and k of this rule are called **parameters** of the family; they are arbitrary constants, except that $a \neq 0$.

The specific function defined by $f(x) = |x|$ is a member of the family—the member obtained by choosing $a = 1$, $h = 0$, and $k = 0$. The specific function defined by $g(x) = -3|x + 2| + 5$ is also a member of the family—the member obtained by choosing $a = -3$, $h = -2$, and $k = 5$.

We are now ready to make some general observations about this family of functions:

(1) *The graph of each member of the family is an angle.* The rule $f(x) = a|x - h| + k$ can be restated in two-part form:

$$f(x) = \begin{cases} a(x - h) + k, & \text{if } x \geqq h \\ a(h - x) + k, & \text{if } x \leqq h \end{cases}$$

The graph of each part is a ray; the slope of one ray is a, and the slope of the other is $-a$. Notice that substitution of $x = h$ in either part of the rule yields $f(h) = k$.

(2) *The graph of each member of the family is symmetric to the line*

$\{(x, y): x = h\}$. The reader is asked to prove this in Exercise 1.

(3) *If $a > 0$, the range of a member of this family of functions is $\{y: y \geqq k\}$; if $a < 0$, the range of a member of this family of functions is* $\{y: y \leqq k\}$. If $a > 0$, then $a|x - h|$ is positive or 0 for every choice of x; so k is the *minimum* number in the range, and the graph opens upward. On the other hand, if $a < 0$, then $a|x - h|$ is negative or 0 for every choice of x; k is the *maximum* value in the range; and the graph opens downward.

(4) *The vertex of the graph of each member is at (h, k) for the specific values of h and k in the expression for that member.* This is the common endpoint of the two rays that form the graph, and it is also on the axis of symmetry.

(5) *The size of the angle that forms the graph of a member is governed by $|a|$.* Let $f(x) = a|x - h| + k$ and $g(x) = a'|x - h'| + k'$ be two different members of the family. The graphs of f and g are congruent if and only if $|a| = |a'|$. This can be verified intuitively by observing that if $|a| = |a'|$ then the graph of g can be obtained from the graph of f by a sequence of rigid motions; but if $|a| \neq |a'|$, something other than a rigid motion is required to force one figure to fit the other.

It is also worth noting that each member of the family has exactly two zeros or one or none. The three possibilities are illustrated in Figure 4-12. Moreover, if c denotes an arbitrary real number, then

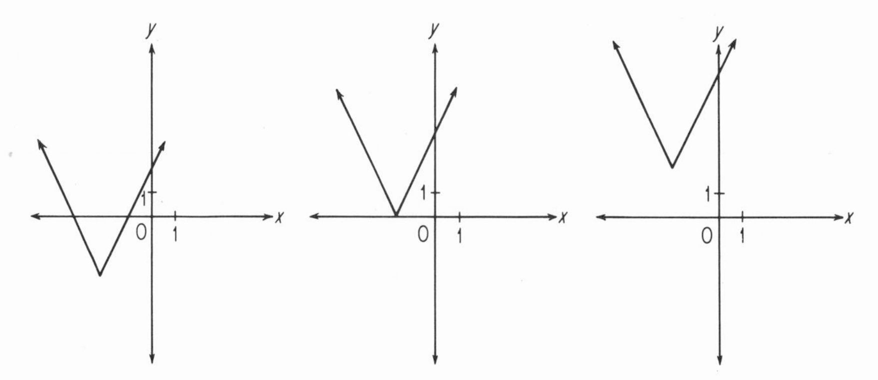

Figure 4-12

each member of the family of functions maps exactly two elements onto c, or exactly one element onto c, or exactly no elements onto c.

In order to find the elements in the domain that have a given image it is necessary to solve an equation.

EXAMPLE 1: Let $f(x) = 2|x - 3| - 4$. Sketch the graph of f. Does f have any zeros? The graph of f is an angle with vertex at

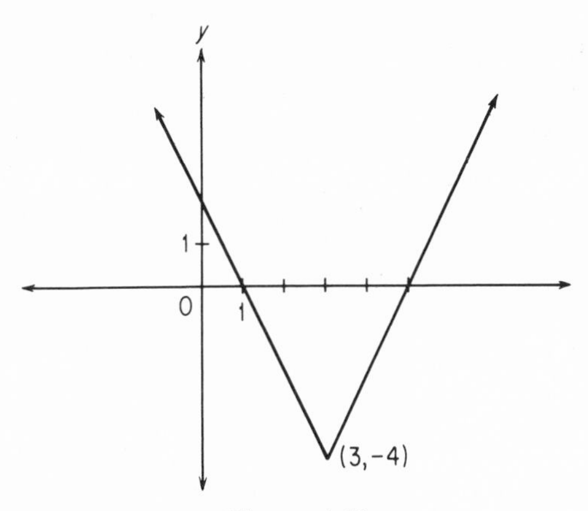

Figure 4-13

$(3, -4)$. The axis of symmetry is $\{(x, y): x = 3\}$. The angle is congruent to the graph of $g(x) = 2|x|$. Furthermore,

$$\begin{aligned} f(x) = 0 &\Leftrightarrow 2|x - 3| - 4 = 0 \\ &\Leftrightarrow 2|x - 3| = 4 \\ &\Leftrightarrow |x - 3| = 2 \\ &\Leftrightarrow x = 1 \quad \text{or} \quad x = 5. \end{aligned}$$

So the zeros are 1 and 5. (See Figure 4-13.)

EXAMPLE 2: Does the function of Example 1 map any numbers onto 50?

$$\begin{aligned} f(x) = 50 &\Leftrightarrow 2|x - 3| - 4 = 50 \\ &\Leftrightarrow |x - 3| = 27 \\ &\Leftrightarrow x = 30 \text{ or } -24 \end{aligned}$$

The numbers 30 and -24 are mapped onto 50. No other numbers are.

EXAMPLE 3: Does the function of Example 1 map any numbers onto -10?

The answer is "no," since the minimum number of the range is -4.

4-4 EXERCISES

1. Prove that the graph of f defined by $f(x) = a|x - h| + k$, $a \neq 0$, is symmetric to the line $\{(x, y): x = h\}$.

2. Sketch the graph of each of these functions. Label the vertex and the axis of symmetry.

(a) $y = 2|x - 1| + 4$ (b) $y = -2|x - 1| + 4$
(c) $y = -2|x - 1| - 4$ (d) $y = \frac{1}{2}|x + 4|$
(e) $y = \frac{1}{2}|x + 4| + 3$ (f) $y = 5|x| + 2$

3. Consider the one-parameter family of functions defined by

$$f(x) = a|x - 2|.$$

On the same set of coordinate axes sketch the graph of the four members of this family obtained by taking $a = -1, 1, 2, 4$.

4. Consider the one-parameter family of functions defined by

$$f(x) = |x - 2| + k.$$

Sketch the graph of the four members obtained by taking $k = -1, 0, 2, 5$.

5. For each part of this problem there is one and only one member of the family $f(x) = a|x - h| + k$ that will meet all the stated conditions. Find that member (that is, assign values to a, h, and k to meet the required conditions).

(a) The graph opens upward, has vertex at (3, 5), and is congruent to the graph of $f(x) = |x|$.
(b) The graph opens upward, has vertex at (0, −2), and is congruent to the graph of $f(x) = 2|x|$.
(c) The graph opens downward, has vertex at (6, 0), and is congruent to the graph of $f(x) = -\frac{1}{2}|x|$.
(d) The number 8 is the minimum number in the range, the axis of symmetry is $\{(x, y): x = 2\}$, and the graph contains the point (3, 10).
(e) The number 12 is the maximum number in the range; the axis of symmetry is $\{(x, y): x = -2\}$, and the graph contains the point (4, 0).

6. (a) Sketch the graph of $f(x) = 3|x - 4| - 2$.
(b) Does this function have any zeros? If so, what are they?
(c) What numbers in the domain map onto 10?
(d) What numbers in the domain map onto 100?

7. (a) Sketch the graph of $g(x) = -|x + 1| - 1$.
(b) Does this function have any zeros? Explain.
(c) What numbers in the domain map onto −2?

8. (a) Sketch the graph of $h(x) = 4 - |x|$.
(b) Does this function have any zeros? If so, what are they?
(c) All numbers on a certain interval of the domain have positive images. Give a concise description of this interval.

9. (a) Sketch the graph of $F(x) = 3|x - 2|$.
(b) Does F have any zeros?
(c) What numbers in the domain have positive images?

10. Let f be defined by $f(x) = |x|$ and g be defined by $g(x) = |x + 3| - 10$. The graphs of these two functions are congruent. Suggest a sequence of reflections of the graph of f that will produce the graph of g.

11. Let f be defined by $f(x) = 2|x - 4| + 3$ and g be defined by $g(x) = -2|x + 6| + 9$. Suggest a sequence of reflections of the graph of f that will produce the graph of g.

5

Quadratic Functions

5-1 DEFINITION AND EXAMPLES

Definition: Let the domain of a function f be the set of real numbers or some non-empty subset of the real numbers. Suppose the rule for f can be put in the form $f(x) = ax^2 + bx + c$, where a, b, and c denote real numbers and where $a \neq 0$. Then f is called a **quadratic function.**

EXAMPLES: Each of these is a rule for a quadratic function:

$$f(x) = 3x^2 + 2x - 1$$
$$g(t) = t^2$$
$$y = 1 - x^2$$
$$y = (x - 1)(x - 2)$$
$$y = (t - 1)^2 + 6$$

In some of these examples the given rule is not in the form $f(x) = ax^2 + bx + c$ specified by the definition. But each of these rules can be put in that form without changing its meaning. For example, the rule

$$y = (x - 1)(x - 2)$$

is the same rule as

$$y = x^2 - 3x + 2$$

because for all real numbers x,

$$(x - 1)(x - 2) = x^2 - 3x + 2.$$

Similarly, the rules

$$y = (t - 1)^2 + 6$$

and

$$y = t^2 - 2t + 7$$

are rules that are different in appearance but not in meaning.

5-1 EXERCISES

Write each of these rules for quadratic functions in the form $y = ax^2 + bx + c$. Identify a, b, and c.

1. $y = (x + 3)^2$

2. $y = (2x - 1)^2 + 5$

3. $y = 2(x + 3)^2 + 7$

4. $y = \pi(x - 1)^2$

5. $y = (x + 1)^2 + (x + 2)^2$

6. $y = (x - 1)^2 - 2(x + 3)^2$

7. $y = (x - 1)(x + 5)$

8. $y = (2x + 1)(x + 7)$

9. $y = (x - 2)(x + 4)$

10. $y = x(x + 2)$

5-2 A REVIEW OF SOME FACTS ABOUT NUMBERS

At this point it is appropriate to discuss a few algebraic facts that the reader must know if he is to handle quadratic functions with ease. Let x, a, b, and c denote arbitrary real numbers.

(1) $ab + ac = a(b + c)$.
(2) $a^2 + 2ab + b^2 = (a + b)^2$.
(3) $a^2 - 2ab + b^2 = (a - b)^2$.
(4) $a^2 - b^2 = (a - b)(a + b)$.
(5) $x^2 - (a + b)x + ab = (x - a)(x - b)$.

The first of these is the distributive law—one of the basic properties of the real number system. The other four generalizations can be proved by using the distributive law and other basic properties of numbers (Section 3-4). The student is asked to provide these proofs (Exercise 4 following this section).

These five statements are about numbers and it is very important that the reader understand this—and also understand that each one asserts infinitely many numerical facts. For example, the second statement says (in part) that

$$(3 + 47)^2 = 3^2 + 2 \cdot 3 \cdot 47 + 47^2$$

One can rely on this equality without doing the arithmetic. But of course the generalization

$$(a + b)^2 = a^2 + 2ab + b^2$$

says much more than this fact about $(3 + 47)^2$; the assertion is that

$$(a + b)^2 = a^2 + 2ab + b^2$$

no matter what numbers are denoted by a and b.

Moreover, the choice of the letters a, b, c, and x to denote numbers is entirely arbitrary; other symbols could serve just as well. To say that

$$(s + t)^2 = s^2 + 2st + t^2$$

for all numbers s and t is simply to repeat that

$$(a + b)^2 = a^2 + 2ab + b^2$$

for all numbers a and b. And if we say that

$$(2x + 3y)^2 = (2x)^2 + 2(2x)(3y) + (3y)^2$$

this is still the same statement with the numbers now called $2x$ and $3y$ instead of a and b.

One other observation should be made about the five generalizations listed in this section: In each case the form on the right of the equation is called a **factored** form of the expression—meaning that the expression is written as a product. Just as 15 can be written in the form $3 \cdot 5$, so can $x^2 + 3x - 4$ be written in the factored form $(x + 4)(x - 1)$ because for every x,

$$x^2 + 3x - 4 = (x + 4)(x - 1).$$

Some expressions used to define quadratic functions can be factored (over the real numbers) and others cannot. The rule

$$f(x) = x^2 - 5x - 6$$

is the same rule as

$$f(x) = (x - 6)(x + 1)$$

because

$$x^2 - 5x - 6 = (x - 6)(x + 1)$$

for all numbers x. Likewise,

$$g(x) = x^2 + 8x + 16$$

is the same rule as

$$g(x) = (x + 4)^2$$

and

$$h(x) = 2x^2 + 8x$$

is the same rule as

$$h(x) = 2x(x + 4).$$

By definition a quadratic function rule is one that can be put in the form

$$f(x) = ax^2 + bx + c.$$

Sometimes the rule can also be written in **factored** form:

$$f(x) = a(x - r)(x - s)$$

for some choice of real numbers r and s. If $ax^2 + bx + c$ can be factored, then the factors are unique, as we observed in Section 2-4. A discussion of the conditions that insure that there is a factored form of the rule is given in Section 5-6. Meanwhile, the reader needs skill in changing the rule from one form to the other if there is a factored form.

5-2 EXERCISES

1. Each of these rules for a quadratic function is presented in factored form. Change each to the form $y = ax^2 + bx + c$.

(a) $y = (x - 2)(x + 3)$ (b) $y = -4(x + 1)^2$
(c) $y = -(x - 5)^2$ (d) $y = 3(x + 2)^2$
(e) $y = (x + 3)(x - 3)$ (f) $y = 4(x - 5)(x + 5)$
(g) $y = (2 - x)(3 + x)$ (h) $y = (4 - x)(x + 5)$
(i) $y = 10(x - \pi)(x + \pi)$ (j) $y = \pi(x + 7)^2$
(k) $y = x(x + 4)$ (l) $y = 3x(x - 5)$
(m) $y = 7(x + 3)(x - 2)$ (n) $y = -2(x + 4)(x - 3)$

2. Each of these rules is presented in the form $y = ax^2 + bx + c$. Write the rule in the form $y = a(x - r)(x - s)$. (Note: These examples have been selected so that this can be done.)

(a) $y = x^2 + 10x + 25$ (b) $y = x^2 - 3x - 4$
(c) $y = 2x^2 + 3x + 1$ (d) $y = 2x^2 + 8x$
(e) $y = 7x^2 - x$ (f) $y = x^2 - 25$
(g) $y = 2x^2 - 50$ (h) $y = \frac{1}{4}x^2 - 25$
(i) $y = -x^2 - 5x + 6$ (j) $y = -3x^2 + 8x$

3. The principal advantage in having a quadratic function rule given in factored form is that in this form the zeros of the function can be obtained at a glance. (A product of two numbers is 0 if and only if one of the factors is zero.) Find all the zeros of the functions in Exercise 2 without doing any additional pencil work.

4. Use the basic properties of real numbers to prove statements (2), (3), (4), and (5) of this section.

5-3 ANOTHER FORM FOR THE RULE OF A QUADRATIC FUNCTION

The idea of changing the form of a function rule without changing its meaning is an important one. The theorem that follows is about such a change in form.

Theorem 5-1: Let f be a function with rule $f(x) = ax^2 + bx + c$, where $a \neq 0$. Then the rule for f can be written in the form

$$f(x) = a(x - h)^2 + k$$

where

$$h = \frac{-b}{2a} \quad \text{and} \quad k = c - \frac{b^2}{4a}.$$

Proof: For every real number x

$$\begin{aligned} a\left[x - \left(-\frac{b}{2a}\right)\right]^2 + c - \frac{b^2}{4a} &= a\left(x + \frac{b}{2a}\right)^2 + c - \frac{b^2}{4a} \\ &= a\left(x^2 + \frac{b}{a}x + \frac{b^2}{4a^2}\right) + c - \frac{b^2}{4a} \\ &= ax^2 + bx + c. \end{aligned}$$

The significance of this theorem will become apparent when the graph of a quadratic function is discussed in the next section.

In applying the theorem to specific functions the student is not required to memorize the formula that expresses k in terms of a, b, and c. It *is* a good idea to remember that $h = (-b)/2a$; then k can be calculated to fit the specific function.

EXAMPLES: Convert each of these function rules to the form $y = a(x - h)^2 + k$.

(1)

$$\begin{aligned} y &= x^2 - 8x + 1 \\ &= (x - 4)^2 + k \qquad \left(\text{since } h = \frac{-(-8)}{2}\right) \\ &= (x - 4)^2 - 15 \qquad (\text{solve } 16 + k = 1) \end{aligned}$$

Thus $a = 1$, $h = 4$, and $k = -15$.

(2) $y = 2x^2 + 8x$

$$= 2[x - (-2)]^2 + k \qquad \left(\text{since } h = \frac{-8}{4}\right)$$

$$= 2[x - (-2)]^2 - 8 \qquad [\text{solve } 2(-2)^2 + k = 0]$$

In this example $a = 2$, $h = -2$, and $k = -8$.

(3) $y = -2x^2 - 12x + 5$

$$= -2[x - (-3)]^2 + k \qquad \left(\text{since } h = \frac{12}{-4}\right)$$

$$= -2[x - (-3)]^2 + 23 \qquad [\text{solve } -2\cdot(-3)^2 + k = 5]$$

Here $a = -2$, $h = -3$, and $k = 23$.

5-3 EXERCISES

1. Write each of these rules in the form $y = ax^2 + bx + c$.

(a) $y = (x + 3)^2 - 5$ (b) $y = 2(x - 1)^2$
(c) $y = (x + 7)^2 + 10$ (d) $y = -4(x + 2)^2 + 7$
(e) $y = (1 - x)^2 + 3$ (f) $y = (x + 1)^2 + (x + 2)^2$
(g) $y = x^2 + (2x - 1)^2$ (h) $y = 7(x + 3)^2 - 40$
(i) $y = x(x - 8)$
(j) $y = (x - 1)^2 + (x - 2)^2 + (x - 4)^2$

2. Write each of these rules in the form $y = a(x - h)^2 + k$. Identify a, h, and k.

(a) $y = x^2 + 6x$ (b) $y = x^2 - 6x$
(c) $y = x^2 + 8x + 5$ (d) $y = x^2 - 2x - 1$
(e) $y = 4x^2 + 8x$ (f) $y = -4x^2 + 8x + 1$
(g) $y = \frac{1}{2}x^2 + 6x$ (h) $y = -\frac{1}{2}x^2 - 6x$
(i) $y = 2x^2$ (j) $y = (x + 1)^2$

5-4 GRAPHING QUADRATIC FUNCTIONS

The rules

$$f(x) = a|x - h| + k$$

$$g(x) = a(x - h)^2 + k$$

are so similar in form that the reader naturally expects the function f (for a specific choice of a, h, and k) to have some properties in common with the function g (for the same choice of a, h, and k). The two do have much in common, and the reader who knows how to graph the absolute value functions of Chapter 4 will find it easy to graph the quadratic functions of Chapter 5.

As a first example, compare f defined by

$$f(x) = |x|$$

to g defined by

$$g(x) = x^2.$$

Observe that f maps x onto the distance from 0 to x, and g maps x onto the *square* of the distance from 0 to x. The reader can verify that the graphs of these two functions (Figure 5-1) have the same axis of

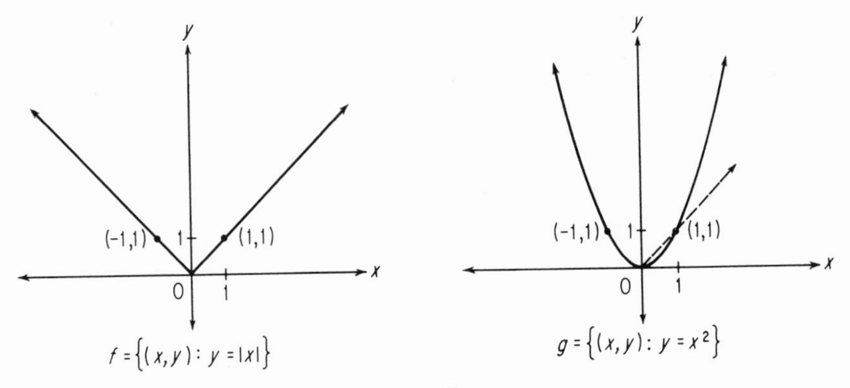

Figure 5-1

symmetry (the y-axis), the same range ($\{y : y \geqq 0\}$), and the same vertex [the point $(0, 0)$]. (In each case the **vertex** is defined as the unique point in which the axis of symmetry intersects the graph of the function.)

Of course, the two graphs are not the same—one is an angle; the other is a curve called a parabola. They have only three points in common. This can be determined by solving the equation $|x| = x^2$. The only solutions to this equation are 0, 1, and -1; therefore, the only points that are common to the graph of $\{(x, y) : y = |x|\}$ and the graph of $\{(x, y) : y = x^2\}$ are the points with coordinates $(0, 0)$, $(1, 1)$, and $(-1, 1)$. If $|x| < 1$, then $x^2 < |x|$, so on the interval where $|x| < 1$ the parabola lies "below" or on the angle. On the other hand, if $|x| > 1$, then $x^2 > |x|$; on the interval where $x > 1$ and the interval where $x < -1$ the parabola lies "above" the angle. As the values of x increase $(x > 1)$, the distance between $(x, f(x))$ on the angle and $(x, g(x))$ on the parabola becomes greater and greater. Intuitively it is also clear that the parabola, like the angle, is a smooth graph with no gaps.

As a second example, compare the two functions f and g, where

$$f(x) = 2|x - 5| + 3$$
$$g(x) = 2(x - 5)^2 + 3.$$

Observe that f maps x onto 3 more than twice the distance from 5 to x, and g maps x onto 3 more than twice the square of the distance from 5 to x. The graphs of these two functions (Figure 5-2) have the

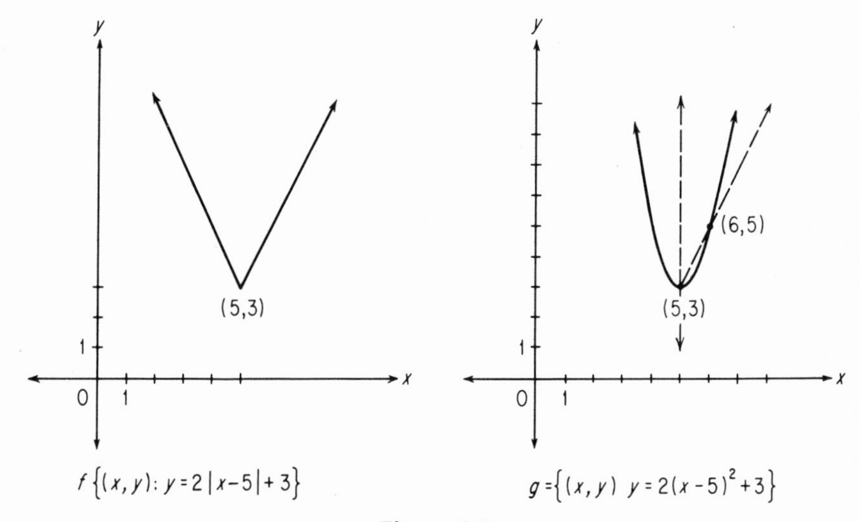

Figure 5-2

same axis of symmetry (the line $\{(x, y): x = 5\}$), the same range $(\{y: y \geqq 3\})$, and the same vertex [the point (5, 3)]. They have in common the vertex and two other points: (6, 5) and (4, 5).

Now let f denote an arbitrary quadratic function. Its graph is called a **parabola**; a few observations are enough to determine what it looks like. By Theorem 5-1 the rule for f can be put in the form $f(x) = a(x - h)^2 + k$, where $a \neq 0$. Using this form of the rule, the reader can prove three important generalizations. A fourth is also listed below, even though we cannot prove it because it concerns congruence—an idea that has not been precisely defined in this book.

(1) The graph of f is symmetric with respect to $\{(x, y): x = h\}$.

(2) If $a > 0$, the range of f is $\{y: y \geqq k\}$; the parabola opens upward, and k is the **minimum number in the range.** If $a < 0$, the range of f is $\{y: y \leqq k\}$; the parabola opens downward, and k is the **maximum number in the range.**

(3) The point (h, k) is on the graph and on the axis of symmetry. (It is called the **vertex.**)

(4) Let g be a second quadratic function, defined by

$$g(x) = a'(x - h')^2 + k'.$$

The graph of f is congruent to the graph of g if and only if $|a| = |a'|$.

Notice that these four generalizations are similar to statements made about the family of absolute value functions discussed in Chapter 4.

With a knowledge of these facts the reader can quickly make a sketch of any quadratic function. Remember that the parameters h and k determine the location of the vertex of the parabola; the parameter a governs the size and determines whether the parabola opens upward or downward.

EXAMPLE 1: Sketch the graph of f defined by

$$f(x) = x^2 + 6x.$$

First change the form of the rule to

$$f(x) = [x - (-3)]^2 - 9.$$

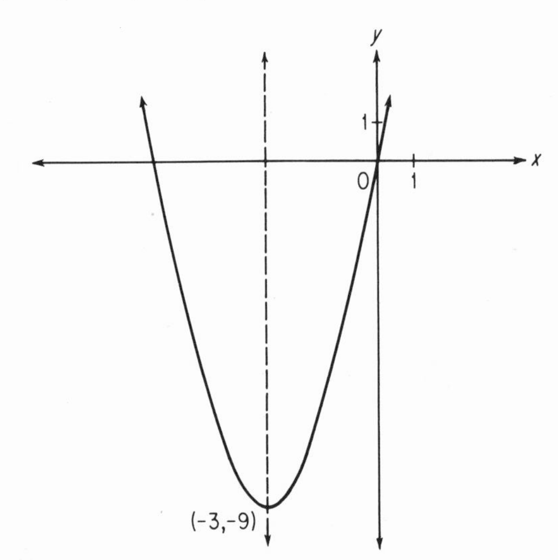

Figure 5-3

The graph (Figure 5-3) is a parabola with vertex at $(-3, -9)$; it opens upward and is congruent to the graph of g defined by $g(x) = x^2$.

EXAMPLE 2: Sketch the graph of f defined by

$$f(x) = -\tfrac{1}{2}x^2 + 4x - 5.$$

This rule is the same as

$$f(x) = -\tfrac{1}{2}(x - 4)^2 + 3.$$

The graph (Figure 5-4) is a parabola with vertex at $(4, 3)$; it opens downward and is congruent to the graph of g defined by $g(x) = \frac{1}{2}x^2$.

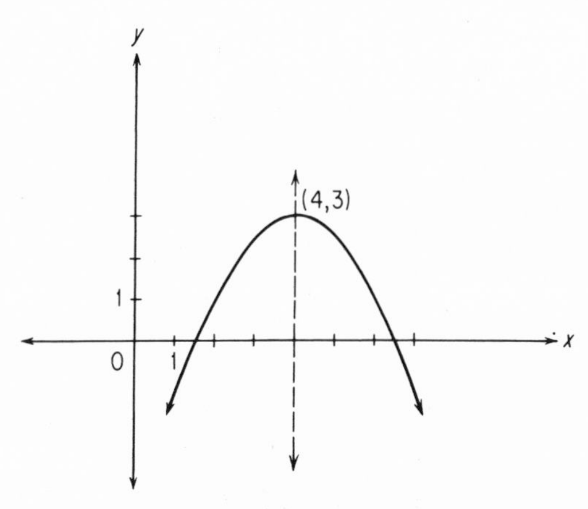

Figure 5-4

EXAMPLE 3: Sketch the graph of several members of the one-parameter family defined by

$$f(x) = a(x - 2)^2.$$

All the parabolas of this family have the same vertex $(2, 0)$ and The same axis of symmetry, but they are not all congruent depending on the choice of a. (See Figure 5-5.)

EXAMPLE 4: Sketch the graph of several members of the one-parameter family defined by

$$f(x) = \tfrac{1}{2}(x - 2)^2 + k.$$

All the parabolas of this family are congruent and have the same axis of symmetry, but they have different vertices (Figure 5-6).

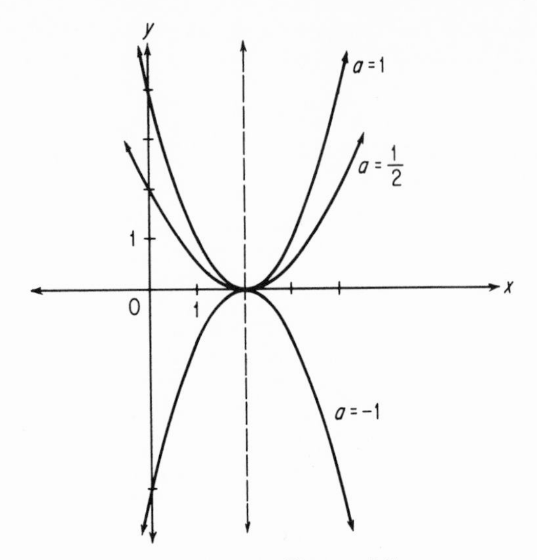

Figure 5-5

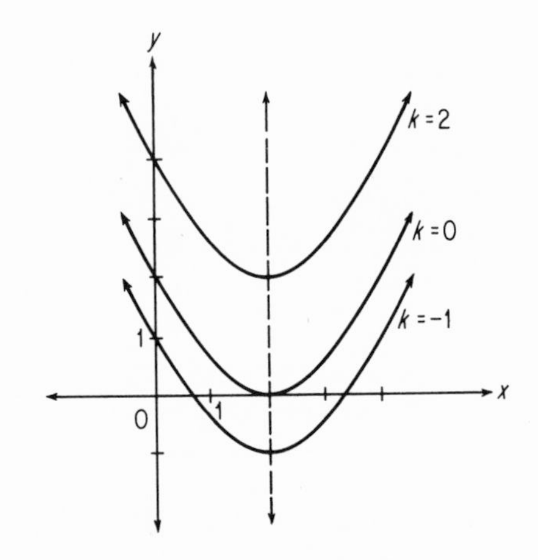

Figure 5-6

EXAMPLE 5: Write the rule for a quadratic function if its graph has its vertex at (1, 2) and is congruent to the parabola defined by

$$f(x) = 2x^2 + 5.$$

First write the rule for the family of parabolas with vertex (1, 2). The equation $g(x) = a(x - 1)^2 + 2$, $a \neq 0$, describes this family. A member of this family is congruent to the graph of $f(x) = 2x^2 + 5$ if and only if $|a| = 2$; that is, if and only if $a = 2$

or $a = -2$. Therefore, there are two different quadratic functions that satisfy the required conditions:

$$g(x) = 2(x - 1)^2 + 2$$
$$h(x) = -2(x - 1)^2 + 2.$$

5-4 EXERCISES

1. In the same coordinate plane sketch:
(a) $f(x) = |x| + 2$ and $g(x) = x^2 + 2$.
(b) $f(x) = 2|x - 3|$ and $g(x) = 2(x - 3)^2$.
(c) $f(x) = \frac{1}{2}|x - 1|$ and $g(x) = \frac{1}{2}(x - 1)^2$.
(d) $f(x) = -\frac{1}{2}|x - 2| + 1$ and $g(x) = -\frac{1}{2}(x - 2)^2 + 1$.

2. Sketch the graph of each of these quadratic functions. Label the coordinates of the vertex of the parabola, and state specifically what the axis of symmetry is.
(a) $y = x^2 + 5$
(b) $y = (x - 1)^2 + 5$
(c) $y = 2(x + 1)^2 - 4$
(d) $y = -2(x + 1)^2 - 4$
(e) $y = x^2 + 6x$
(f) $y = x^2 + x + 1$
(g) $y = -(x - 4)^2$
(h) $y = x^2 - 4$
(i) $y = 2x^2 + 8x + 1$
(j) $y = 1 + 2x + x^2$

3. The range of each function in Exercise 2 contains either a maximum or minimum number. Find this number and state whether it is a maximum or minimum.

4. The equation $f(x) = a(x + 2)^2 + 1$ defines a one-parameter family of functions. On the same set of axes sketch the four members of this family that are obtained by taking $a = -1, 0, 1, 4$.

5. The equation $f(x) = (x + 2)^2 + k$ defines a one-parameter family of functions. On the same set of axes sketch the four members of this family that are obtained by taking $k = -1, 0, 1, 4$.

6. The equation $f(x) = -\frac{1}{2}(x - h)^2 + 1$ defines a one-parameter family of functions. On the same set of axes sketch the four members of this family that are obtained by taking $h = -1, 0, 1, 4$.

7. Write the rule for a quadratic function if:
(a) Its graph has vertex at (2, 3) and is congruent to the parabola defined by $g(x) = 4x^2$.
(b) Its graph has vertex at (−1, 6) and is congruent to the parabola defined by $g(x) = -5x^2$.
(c) Its graph has vertex at (0, 0) and contains the point (3, 18).
(d) Its graph has vertex at (2, 5) and contains the point (0, 0).

8. The domain of a quadratic function need not be the entire set of real numbers. Sketch the graph of each of these functions, being sure to note the restrictions on the domain.

(a) $f(x) = x^2 + 2x$, domain: $\{x: x \geqq 0\}$.
(b) $g(x) = x^2$, domain: all integers.
(c) $F(x) = 1 - x^2$, domain: $\{x: x \leqq 0\}$.
(d) $G(x) = (x - 4)^2$, domain: $\{x: x \geqq 0\}$.

†**9.** (a) Draw the graph of $f(x) = x^2 + 2$. Reflect this in the x-axis to get the graph of a function that we shall call g. Then reflect the graph of g in the line $\{(x, y): x = 4\}$ to get the graph of another function that we shall call h.
(b) Write the rules for the two functions—g and h—described in part (a).

†**10.** Let f be defined by $f(x) = a(x - h)^2 + k$, where $a \neq 0$.
(a) Prove that the graph of f is symmetric to the line $\{(x, y): x = h\}$.
(b) Prove that if $a > 0$ the range of f is $\{y: y \geqq k\}$, and if $a < 0$ the range of f is $\{y: y \leqq k\}$. [Hint: For every number x, $(x - h)^2 \geqq 0$, and $(x - h)^2 = 0 \Leftrightarrow x = h$.]

5-5 DOES THE QUADRATIC FUNCTION f MAP ANY NUMBER ONTO c?

Problems of two general types occur again and again in the study of functions. These will be illustrated by considering the function with the set of all real numbers as its domain and rule $f(x) = x^2 - 10$.

Question (Type 1): What number in the range does f map 4 onto? (See Figure 5-7.)
Answer: $f(4) = 4^2 - 10 = 6$.

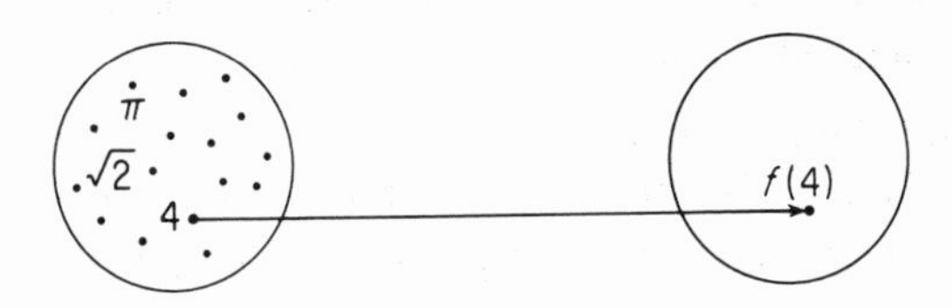

Figure 5-7

Question (Type 2): Is there any element in the domain that f maps onto 15? (See Figure 5-8.)
Answer: If $f(x) = 15$ for some particular x, then $x^2 - 10 = 15$.

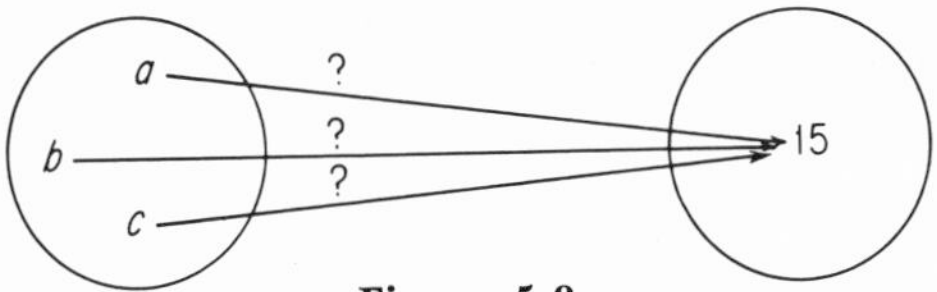

Figure 5-8

This is equivalent to saying $x^2 = 25$, which, in turn, is the same as saying $x = 5$ or -5. So yes, there is an element that f maps onto 15; in fact, there are exactly two of them, and these are 5 and -5.

In this section we shall investigate questions of Type 2. Given a quadratic function f and a real number c, find all the numbers (if any) that are mapped onto c by f.

EXAMPLE 1: Let g have as its domain the set of all real numbers and rule $g(x) = x^2 - x - 2$. Are there any numbers that g maps onto 0?

The rule for g can be written

$$g(x) = (x - 2)(x + 1).$$

A fundamental property of numbers is that *the product of two numbers is 0 if and only if one (or both) of the numbers is 0.* Applying this property, we see that $g(2) = 0$ and $g(-1) = 0$; moreover, since factorization is unique, these are the only two numbers that g maps onto 0. Thus 2 and -1 are the zeros of g. (See Figure 5-9.)

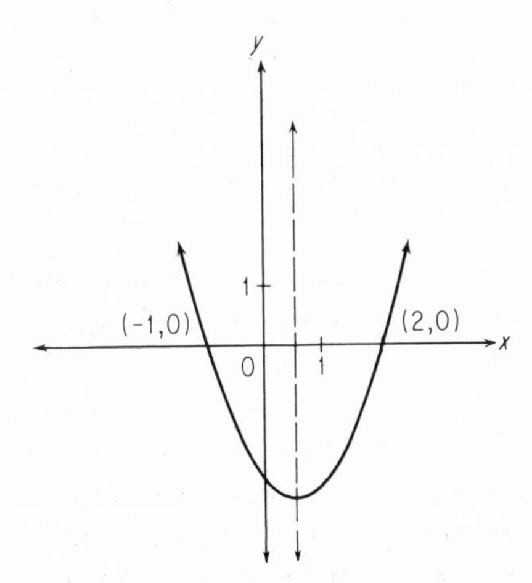

Figure 5-9

EXAMPLE 2: Let $g(x) = x^2 - 2x$. Does g map any numbers onto 15? (See Figure 5-10.)

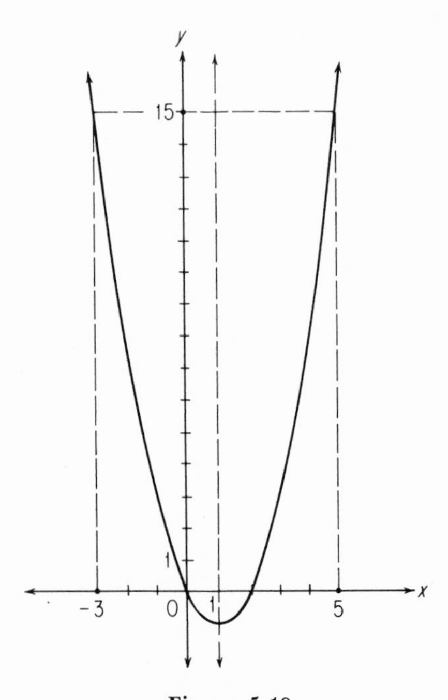

Figure 5-10

$$\begin{aligned} g(x) = 15 &\Leftrightarrow x^2 - 2x = 15 \\ &\Leftrightarrow x^2 - 2x - 15 = 0 \\ &\Leftrightarrow (x - 5)(x + 3) = 0 \\ &\Leftrightarrow x = 5 \quad \text{or} \quad x = -3. \end{aligned}$$

So 5 and -3 are the only numbers mapped onto 15 by g.

Notice that in Example 2 we were not asked to find the *zeros* of g; yet to answer the question we did have to find the zeros of h defined by $h(x) = x^2 - 2x - 15$.

Let $f(x) = x^2 + 7x$; we wish to know the numbers that f maps onto 8. This is equivalent to finding the zeros of $g(x) = x^2 + 7x - 8$. We can answer the original question *provided we can get the zeros of* g.

Suppose that f is any quadratic function, and the problem is to find the numbers that f maps onto some number c. We can solve this problem provided we can get the zeros of the function g defined by $g(x) = f(x) - c$. But since f is quadratic, so is g. If we only knew how to get the zeros of every quadratic function, we would also know how to find all the numbers that f maps onto c. This general problem will be discussed in the next section.

5-5 EXERCISES

1. Sketch the graph of f defined by $f(x) = x^2 + 2x$. Find all the numbers (if any) that f maps onto:

(a) 0
(b) 3
(c) -1
(d) -4 [Hint: Find the minimum number of the range of f.]

2. Sketch the graph of g defined by $g(x) = 4x - x^2$. Find all the numbers (if any) that g maps onto:

(a) 0
(b) 10
(c) -5
(d) -45

3. Find the zeros of the functions described below.

(a) $f(x) = x^2 - 7x - 8$
(b) $g(x) = x^2 + 8x + 16$
(c) $h(x) = x - 5x^2$
(d) $F(x) = x^2 - 5$
(e) $G(x) = (x - 2)^2 - 9$
(f) $H(x) = 2x^2 - x - 1$

5-6 DOES THE QUADRATIC FUNCTION f HAVE ANY ZEROS? IF SO, WHAT ARE THEY?

The first thing to observe is that the number of zeros for a given quadratic function *will be two or one or none*. This is evident from the graph. The three possibilities are illustrated in Figure 5-11.

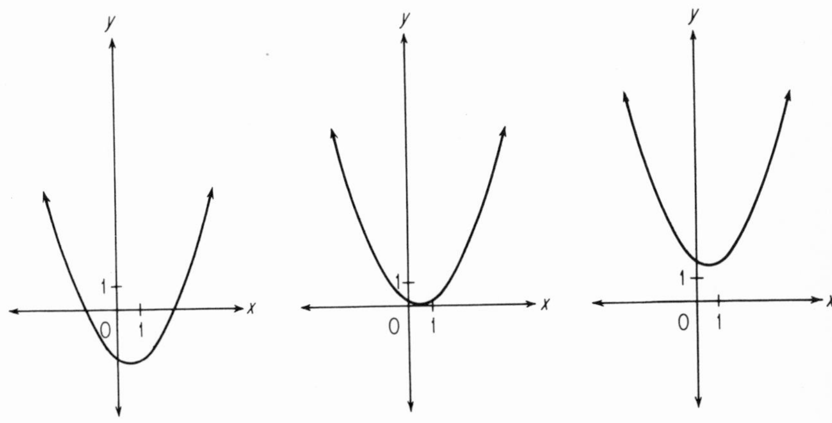

Figure 5-11

In this section it is our purpose to examine these three possibilities in more detail. An earlier theorem stated that the rule $f(x) = ax^2 + bx + c$, where $a \neq 0$, could be written in the form

$$f(x) = a\left[x - \left(-\frac{b}{2a}\right)\right]^2 + \left(c - \frac{b^2}{4a}\right).$$

So

$$f(x) = 0 \Leftrightarrow a\left(x + \frac{b}{2a}\right)^2 + \left(c - \frac{b^2}{4a}\right) = 0$$

$$\Leftrightarrow \left(x + \frac{b}{2a}\right)^2 = \frac{b^2 - 4ac}{4a^2}.$$

The denominator of the fraction on the right is always positive, so the fraction has the same sign as the numerator, $b^2 - 4ac$. Moreover, no choice of x will make $[x + (b/2a)]^2$ negative, and only one choice, $x = -b/2a$, will make $[x + (b/2a)]^2 = 0$. Thus the number of zeros of f depends upon $b^2 - 4ac$. Putting all these facts together, we arrive at an important general conclusion.

Theorem 5-2: Consider a quadratic function f with domain the set of all real numbers and rule

$$f(x) = ax^2 + bx + c,$$

where $a \neq 0$.

1. If $b^2 - 4ac < 0$, then f has no real zeros.
2. If $b^2 - 4ac = 0$, then f has the unique zero $-b/2a$.
3. If $b^2 - 4ac > 0$, then f has two distinct zeros

$$\frac{-b + \sqrt{b^2 - 4ac}}{2a} \text{ and } \frac{-b - \sqrt{b^2 - 4ac}}{2a}.$$

EXAMPLE 1: Let f be defined by

$$f(x) = x^2 + x + 1.$$

For this function (Figure 5-12) $b^2 - 4ac = -3$, so f has no zeros. (As a matter of fact, the minimum number in the range is $\frac{3}{4}$.)

EXAMPLE 2: Let g be defined by

$$g(x) = x^2 + 4x + 4.$$

Since $b^2 - 4ac = 0$, g has a unique zero. This zero is -2. (See Figure 5-13.)

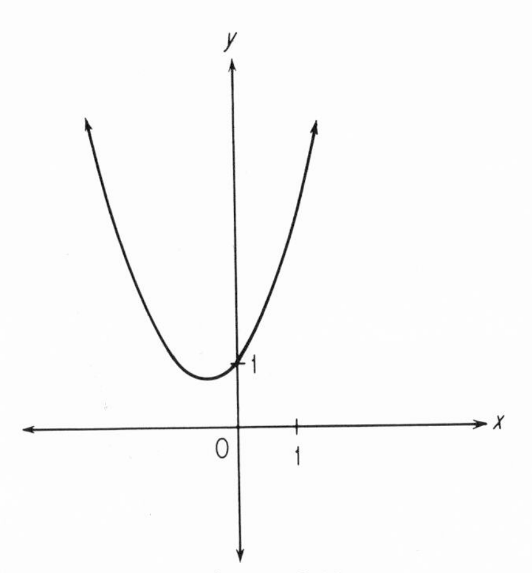

Figure 5-12

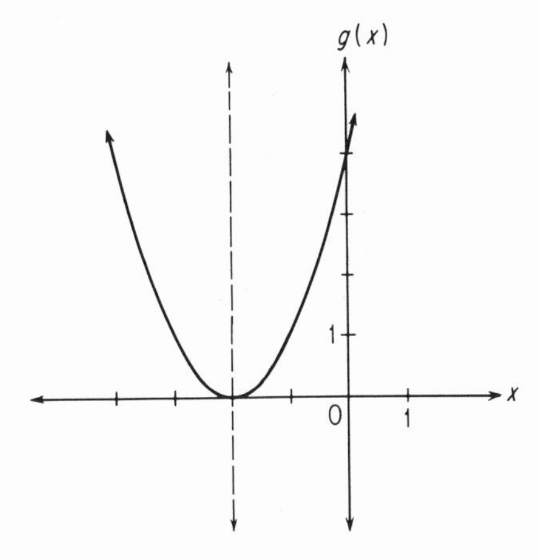

Figure 5-13

EXAMPLE 3: Let h be defined by

$$h(x) = 2x^2 + 5x + 1.$$

Since $b^2 - 4ac = 17$, h has two zeros. These zeros are $(-5 + \sqrt{17})/4$ and $(-5 - \sqrt{17})/4$. (See Figure 5-14.)

Earlier in the chapter it was stated that the rule

$$f(x) = ax^2 + bx + c,$$

where $a \neq 0$, can *sometimes* (but not always) be put in the form $f(x) = a(x - r)(x - s)$, where r and s are real numbers. If a particular function f can be written in this form, then

$$f(r) = a(r - r)(r - s) = 0$$

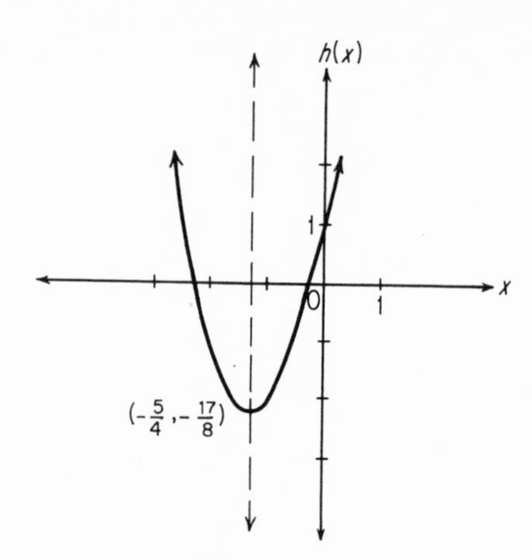

Figure 5-14

and

$$f(s) = a(s - r)(s - s) = 0.$$

This means that the real numbers r and s are zeros of the function. Therefore, $b^2 - 4ac \geqq 0$. [Item (1) of Theorem 5-2 is equivalent to the assertion that if f has real zeros then $b^2 - 4ac \geqq 0$.]

Conversely, suppose

$$f(x) = ax^2 + bx + c$$

where $b^2 - 4ac \geqq 0$. Let

$$r = \frac{-b + \sqrt{b^2 - 4ac}}{2a} \quad \text{and} \quad s = \frac{-b - \sqrt{b^2 - 4ac}}{2a}.$$

Then

$$ax^2 + bx + c = a(x - r)(x - s)$$

as the reader can verify by multiplying the factors on the right (after substituting for r and s) and collecting terms.

Our conclusion is that the function f defined by

$$f(x) = ax^2 + bx + c$$

where $a \neq 0$, can be put in a factored form

$$f(x) = a(x - r)(x - s)$$

where r and s denote real numbers if and only if

$$b^2 - 4ac \geqq 0.$$

5-6 EXERCISES

1. Find all the zeros of these functions, using the easiest possible method.

(a) $y = x^2 - 6x$ (b) $y = 3x^2 - x - 1$

(c) $y = 2x^2 - 8$ (d) $y = x^2 + 5$

(e) $y = x^2 + x - 1$ (f) $y = x^2 + x + 5$

(g) $y = (x - 1)^2$ (h) $y = 3 - x + x^2$

(i) $y = (x - 1)^2 - 4$ (j) $y = 3 - x - 2x^2$

(k) $y = 2x^2 + 3x + 7$ (l) $y = 2x^2 + 3x - 7$

2. Suppose f has as its domain the set of all *positive numbers* and has the rule $f(x) = x^2 + 8x - 9$. What are the zeros of f?

3. Suppose g has as its domain the set of all *positive integers* and has the rule $g(x) = x^2 - 4x - 5$. What are the zeros of g?

4. Suppose h has as its domain the set of all *non-negative integers* and has the rule $h(x) = x^2 + x - 20$. What are the zeros of h?

5. Suppose F has as its domain the set of all *integers* and has the rule $F(x) = 2x^2 + 3x + 1$. What are the zeros of F?

6. Suppose G has as its domain the set of all *positive rational numbers* and has the rule $G(x) = 3x^2 - 8x + 5$. What are the zeros of G?

7. Suppose H has as its domain the set of all *positive real numbers* and has the rule $H(x) = x^2 + \frac{7}{2}x - 2$. What are the zeros of H?

5-7 SOLVING QUADRATIC EQUATIONS

An equation such as

$$t^2 - 7t - 8 = 0$$

is called a **quadratic equation.** Consider the problem of finding all the real numbers that will satisfy the equation:

$$\begin{aligned} & t^2 - 7t - 8 = 0 \\ \Leftrightarrow\ & (t - 8)(t + 1) = 0 \\ \Leftrightarrow\ & t = 8 \quad \text{or} \quad t = -1 \end{aligned}$$

The conclusion is that the equation has exactly two solutions, 8 and -1.

Sometimes, as in the preceding paragraph, a quadratic equation can be "factored" and the solution readily obtained by the use of the law of the zero product. But a more general procedure is needed.

Consider the quadratic equation

$$ax^2 + bx + c = 0$$

where a, b, and c denote arbitrary real numbers, except that $a \neq 0$. What real numbers x will satisfy this equation? This is the same question as: What are the *zeros* of the function that has the set of real numbers as its domain and has the rule

$$f(x) = ax^2 + bx + c?$$

This question was answered in the preceding section. So no further work is needed to see that the equation $ax^2 + bx + c = 0$ has:

(1) No real solution if $b^2 - 4ac < 0$.

(2) A unique solution $-b/2a$ if $b^2 - 4ac = 0$.

(3) Two solutions, $(-b+\sqrt{b^2-4ac})/2a$ and $(-b-\sqrt{b^2-4ac})/2a$, if $b^2 - 4ac > 0$.

5-7 EXERCISES

1. Use the shortest method you can find to solve each of these equations.

(a) $t^2 - 7t = 0$
(b) $3t^2 - 4t = 0$
(c) $t^2 + 8t + 16 = 0$
(d) $2t^2 - t - 1 = 0$
(e) $u^2 - 2u - 15 = 0$
(f) $x^2 + 3x + 1 = 0$
(g) $y^2 + y + 1 = 0$
(h) $(t - 1)^2 = 0$
(i) $(t + 2)^2 - 4 = 0$
(j) $(t + 2)^2 + t + 2 = 0$
(k) $(x - 5)^2 + 6(x - 5) - 7 = 0$
(l) $2u^2 + 5u - 3 = 0$

2. Make up a quadratic equation:

(a) That has 2 and -5 as its solutions.
(b) That has -4 as its only solution.
(c) That has no real number as a solution.
(d) That has 0 and 6 as its solutions.
(e) Such that there are two different solutions but with the same absolute value.

3. Consider the equation $ax^2 + bx = x^2 - 2$, where a and b are arbitrary real numbers. State a necessary and sufficient condition on a and b so that this equation will have two different solutions.

5-8 SOME QUADRATIC INEQUALITIES

After the zeros of a quadratic function have been determined, a rough sketch of the graph of the function will indicate those elements in the domain that have positive images and those that have negative images.

For example, if f has the graph shown in Figure 5-15, we can quickly read these facts:

(1) $f(x) = 0$ if and only if $x = 1$ or $x = 3$.
(2) $f(x) > 0$ if and only if $x < 1$ or $x > 3$.
(3) $f(x) < 0$ if and only if x is between 1 and 3.

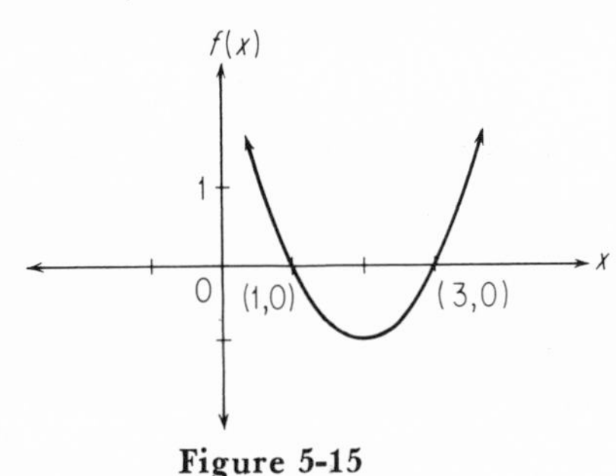

Figure 5-15

5-8 EXERCISES

1. Suppose g has domain all real numbers and rule $g(x) = x^2 - 9x$.
(a) What are the zeros of g?
(b) For what elements in the domain is $g(x)$ positive?
(c) For what elements in the domain is $g(x)$ negative?
(d) For what elements in the domain is $g(x) < 10$?
(e) For what elements in the domain is $g(x) < 22$?
(f) For what elements in the domain is $g(x) < -20$?

2. Make up a rule for some quadratic function f that has the following property: $f(x)$ is positive for every x between 6 and 8, and $f(x)$ is negative or 0 for all other x.

3. Suppose F is defined by the rule $F(x) = (4 - x)(3 + x)$.
(a) What are the zeros of F?
(b) What numbers have negative images?
(c) What numbers have positive images?

4. Suppose G is defined by the rule $G(x) = x^2 + 2$.
   (a) Does G have any zeros?
   (b) What numbers have negative images?
   (c) What numbers have positive images?

5-9 POLYNOMIAL FUNCTIONS

Let the domain of a function f be the set of real numbers or some non-empty subset of the real numbers. Then

$$f(x) = mx + b, \qquad m \neq 0$$

defines a *linear* function, and

$$f(x) = ax^2 + bx + c, \qquad a \neq 0$$

defines a *quadratic* function. The next step up this ladder,

$$f(x) = ax^3 + bx^2 + cx + d, \qquad a \neq 0$$

defines a *cubic* function.

In general, if we let n denote an arbitrary positive integer and $a_0, a_1, \ldots, a_n$ denote arbitrary real numbers with the sole restriction that $a_0 \neq 0$, then the rule

$$f(x) = a_0x^n + a_1x^{n-1} + \cdots + a_{n-1}x + a_n$$

defines a *polynomial function of degree n*. [Also, a constant function, $f(x) = c$, where $c \neq 0$, is called a polynomial function of degree 0.]

Thus the family of polynomial functions is a large family, containing linear functions and quadratic functions as special subfamilies. The study of the properties of polynomial functions in general is too long and difficult to be included in this book. The interested reader is referred to [1], [4] and [8] in the bibliography.

5-10 SOME APPLICATIONS

Quadratic functions are not as widely used in everyday situations as are linear functions, but there are applications in almost every branch of knowledge that makes use of mathematics.

For example, a fundamental problem in physics concerns the motion of an object under the influence of gravity alone. If an object could be hurled straight up from the ground and there were no air resistance, then its height h measured in feet after t seconds would be given by the function rule $h(t) = v_0t - \frac{1}{2}gt^2$, where g denotes the constant acceleration of gravity and v_0 denotes the initial velocity of the object. An approximation to g is 32 ft/sec; after this is substituted, the function rule is $h(t) = v_0t - 16t^2$. Now suppose that an object is hurled vertically upward from the ground with an initial velocity of 96 ft/sec. What is the maximum height it will reach? When will it hit the ground?

The graph of the appropriate function h, which has the rule $h(t) = 96t - 16t^2$, is part of a parabola. Using methods already developed, we find that this parabola opens downward (Figure 5-16)

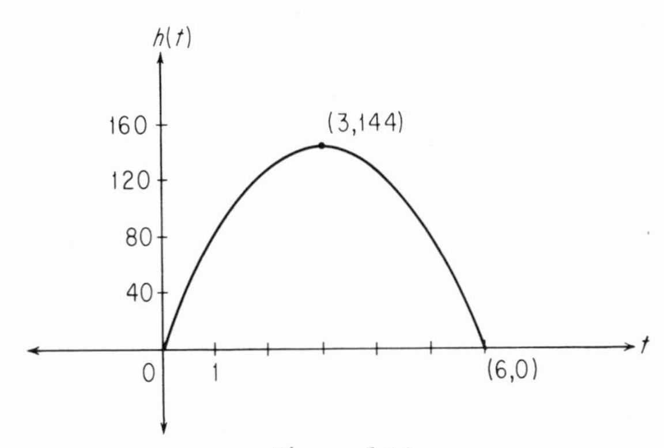

Figure 5-16

and has its vertex at the point (3, 144). Therefore, the maximum height reached by the object is 144 feet. To determine when the object hits the ground, set $h(t) = 0$ and solve for t:

$$\begin{aligned} h(t) = 0 &\Leftrightarrow 96t - 16t^2 = 0 \\ &\Leftrightarrow 16t(6 - t) = 0 \\ &\Leftrightarrow t = 0 \quad \text{or} \quad t = 6. \end{aligned}$$

The object starts upward when $t = 0$; it hits the ground when $t = 6$.

In most applications of quadratic functions the reader will find that he needs to do one (and possibly all three) of the following:

(1) Describe a function that fits the situation outlined in the problem.

(2) Find the maximum or the minimum (if there *is* a maximum or minimum).

(3) Answer questions: Given any specific number c, are there

numbers in the domain of the function that the function maps onto c? If so, what number(s)?

In the foregoing example about an object hurled upward, the reader was *given* the appropriate function—but this is not so in all the exercises that follow. Sometimes the hardest part of the problem is to describe the appropriate function. Remember that in making the description you do not have to use the letters f and x. Often it helps to use letters that suggest the thing being denoted, such as h for height, A for area, etc.

If the appropriate function is quadratic, you should now be able to answer the questions that are asked about it.

5-10 EXERCISES

1. A body is thrown straight up from the ground with an initial velocity of 144 ft/sec. If air resistance is neglected, its height h, measured in feet, after t seconds is given by the formula $h = 144t - 16t^2$.

(a) After how many seconds does the body reach its maximum height?

(b) What is the maximum height reached?

(c) When does the body hit the ground?

2. A projectile is thrown straight up from a height of 6 feet with an initial velocity of 192 ft/sec. If air resistance is neglected, its height h, measured in feet, after t seconds is given by the formula $h = 6 + 192t - 16t^2$. Answer the same three questions as in Exercise 1.

†**3.** Consider the basic function rule of Example 1: $h(t) = v_0t - 16t^2$, where v_0 denotes the initial velocity with which the object leaves the ground. The maximum height reached by the object is clearly a function of v_0.

(a) Describe the function that relates the maximum height reached to the initial velocity.

(b) What initial velocity must be imparted to an object so that the maximum height reached will be 900 feet?

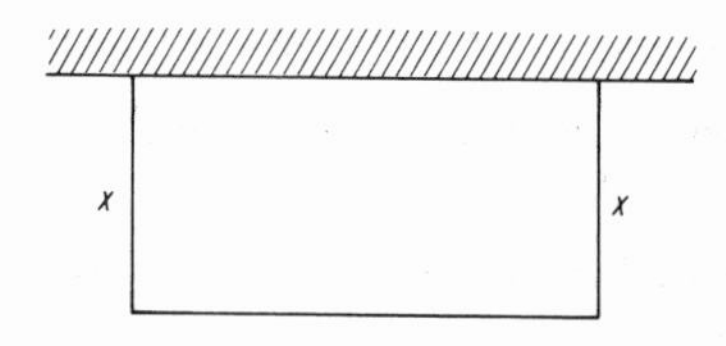

4. A man with 160 feet of fencing wishes to fence off a region in the shape of a rectangle. One side of the region will not require fencing. What should be the dimensions of the rectangle to enclose the largest area possible? [Hint: Write the function rule $A(x)$ that expresses the area in terms of x.]

5. Prove that, of all possible rectangles with a fixed perimeter P, the one that encloses the region with the greatest area is the square.

6. A telephone company can get 1000 subscribers at a monthly rate of \$5.00 each. It will get 100 more subscribers for each 10-cent decrease in the rate. What rate will yield the maximum gross monthly income and what will this income be?

7. A rectangular piece of tin is twice as long as it is wide. From each corner a 2-inch square is cut out, and the ends are turned up so as to make a box whose contents are 60 cubic inches. What are the dimensions of the piece of tin?

8. The edges of a cube are each increased in length by one inch. It is found that the volume of the cube is thereby increased 19 cubic inches. What is the length of the edge of the original cube?

9. Let the radius of a disk be denoted by r, its circumference by C, and its area by A. Recall that $C = 2\pi r$, and $A = \pi r^2$. Suppose the radius of a second disk is three times as long as that of the first.

(a) How do the two circumferences compare?

(b) How do the two areas compare?

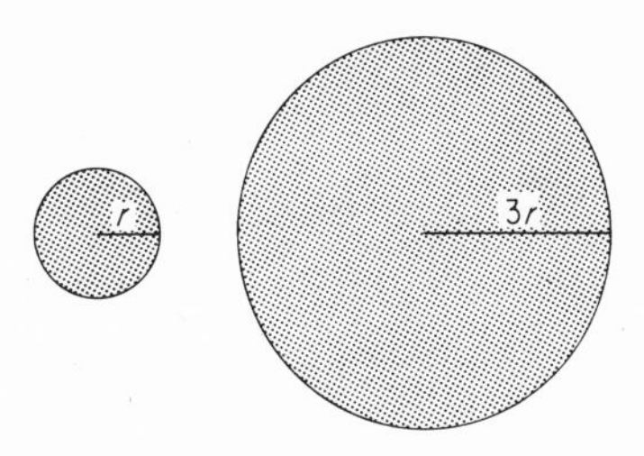

†10. An important problem in statistics is to find the minimum of the quadratic function F defined by $F(x) = (x - a_1)^2 + (x - a_2)^2 + \cdots + (x - a_n)^2$, where $a_1, \ldots, a_n$ are n arbitrary constants.

(a) As a special case, find what choice of x yields the minimum value of $G(x)$, where G is defined by the rule $G(x) = (x - 1)^2 + (x - 3)^2$. [Hint: Recall that $-b/2a$ yields the minimum or maximum of the function with rule $f(x) = ax^2 + bx + c$.]

(b) Now generalize this answer to the function F defined above: Prove that the number in the domain of F that produces the minimum value of $F(x)$ is the *average* of the n constants $a_1, \ldots, a_n$.

6

Using Functions to Construct Other Functions

6-1 ALGEBRAIC OPERATIONS WITH FUNCTIONS

Sometimes two functions can be "added" in a natural way to produce another function. For example, suppose f and g are functions each with the set of all real numbers as its domain and with rules:

$$f(x) = x$$
$$g(x) = |x|.$$

Then $f + g$ is defined to be the function that has the set of all real numbers as its domain and has the rule

$$(f + g)(x) = x + |x|.$$

Thus

$$(f + g)(4) = 8$$
$$(f + g)(-2) = 0$$
$$(f + g)(-5) = 0.$$

In general,

$$(f + g)(x) = \begin{cases} 2x, & \text{if } x \geqq 0 \\ 0, & \text{if } x < 0. \end{cases}$$

The graph of $f + g$ has an obvious geometrical relation to the graphs of f and g. (See Figure 6-1.)

Just as we formed a new function by adding f and g, we might also *subtract* g from f to obtain a new function, $f - g$. Also we might

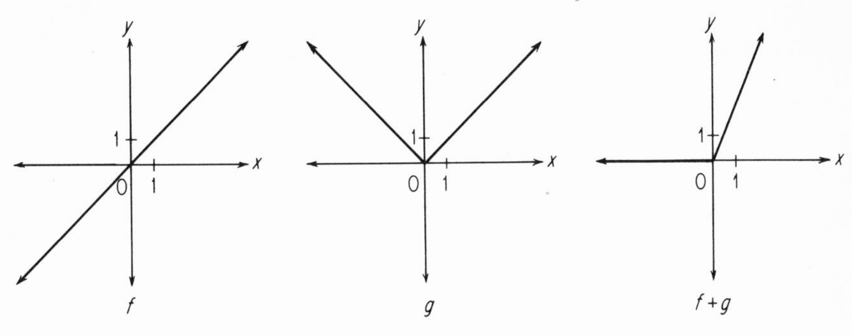

Figure 6-1

multiply or *divide* the two functions. To generalize these ideas and make them precise we need formal definitions.

Definition: Suppose f and g denote functions from a common domain D into the real numbers. Then $f + g$, $f - g$, and $f \cdot g$ are functions with domain D and rules:

$$(f + g)(x) = f(x) + g(x)$$
$$(f - g)(x) = f(x) - g(x)$$
$$(f \cdot g)(x) = f(x) \cdot g(x).$$

Also, f/g is a function that has as its domain all the elements of D for which $g(x) \neq 0$. The rule for f/g is $(f/g)(x) = f(x)/g(x)$. Finally, if c is any real number, $c \cdot f$ is the function with domain D and rule $(c \cdot f)(x) = c \cdot f(x)$.

EXAMPLE: Suppose the domain of f and of g is the set of all real numbers:

$$f(x) = x + 1$$
$$g(x) = x - 2.$$

Then $f + g$, $f - g$, and $f \cdot g$ are functions each with the set of all real numbers as its domain. Their rules are

$$(f + g)(x) = 2x - 1$$
$$(f - g)(x) = 3$$
$$(f \cdot g)(x) = (x + 1)(x - 2).$$

Also, f/g is a function that has the set of all real numbers except 2 as its domain. The rule is

$$\left(\frac{f}{g}\right)(x) = \frac{x+1}{x-2}.$$

Finally, if c denotes an arbitrary real number, then $c \cdot f$ is a function with all real numbers as its domain and with the rule:

$$(c \cdot f)(x) = c(x+1).$$

6-1 EXERCISES

1. Let $f(x) = 2x + 1$ for all x; $g(x) = x + 2$ for all x.
(a) State clearly the domain and the rule for each of these functions: $f+g,\ f-g,\ g-f,\ f \cdot g,\ g \cdot f,\ f/g,\ g/f,\ 7 \cdot f,\ -2g.$

(b) $(f+g)(8) = ?$ $\qquad \dfrac{f}{g}(3) = ?$

$(f \cdot g)(-1) = ?$ $\qquad (3 \cdot f)(0) = ?$

2. Let $f(x) = x$ for all x; $g(x) = |x|$ for all x.
(a) State clearly the domain and rule for each of these functions: $f-g,\ g-f,\ f \cdot g,\ g \cdot f,\ f/g,\ g/f.$
(b) Sketch the graph of each function in part (a).

3. Here is a list of conjectures about functions, some being valid and some invalid. Decide which is which. Give a proof for each of the valid ones and a counterexample for each of the invalid ones. In these statements the letters f, g, and h denote functions with the same domain, each having a subset of the real numbers as its range. You will need to recall the meaning of "equal" functions.
(a) $f + g = g + f.$
(b) $f \cdot g = g \cdot f.$
(c) $f - g = g - f.$
(d) $\dfrac{f}{g} = \dfrac{g}{f}.$
(e) $f + (g + h) = (f + g) + h.$
(f) The sum of any two linear functions is a linear function.
(g) The product of any two linear functions is a quadratic function.
(h) The sum of any two quadratic functions is a quadratic function.

4. The function $(-1 \cdot f)$ is usually written as $(-f)$. Thus, if the rule for f is $f(x) = 3x + 6$, then the rule for $-f$ is $(-f)(x) = -3x - 6$. Draw the graphs of these two functions on the same set of axes.

5. (a) Suppose that the graph of f is a straight line segment as in the figure. Draw the graph of $-f$.

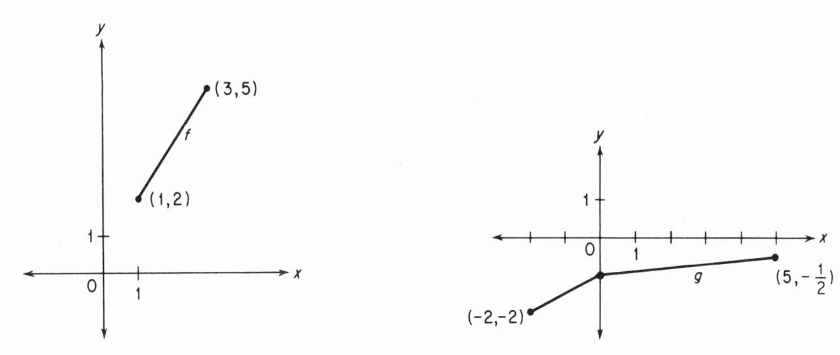

(b) The graph of g is shown above. Draw the graph of $-g$.
(c) Let f denote an arbitrary function from a subset of the real numbers into the real numbers. Is it true that the graphs of f and $-f$ are congruent? Explain.

6. (a) Let

$$f(x) = \frac{1}{x}, \qquad \text{domain: all positive numbers}$$

$$g(x) = 2, \qquad \text{domain: all positive numbers.}$$

Then $(f + g)(x) = (1/x) + 2$, domain: all positive numbers. Draw the graphs of f and $f + g$. Are they congruent graphs? Explain.
(b) Let $g(x) = c$ and suppose that the domain D is a subset of the real numbers. Let f be an arbitrary function from D into the real numbers. Then $f + g$ is defined. Do f and $f + g$ have congruent graphs? Explain.

6-2 RESTRICTIONS AND EXTENSIONS OF FUNCTIONS

A definition from the first chapter is worth repeating here: Two functions f and g are equal if and only if they have the same domain D, and $f(x) = g(x)$ for each x in D.

Thus two functions that have the *same rule* will nevertheless be *different* functions if their domains are different. For example, define two functions f and g as follows:

$$f(x) = 2x + 1, \qquad \text{domain: all real numbers}$$

$$g(x) = 2x + 1, \qquad \text{domain: the non-negative real numbers.}$$

These different functions are obviously related; we call g the **restriction of f to the non-negative numbers**. (See Figure 6-2.)

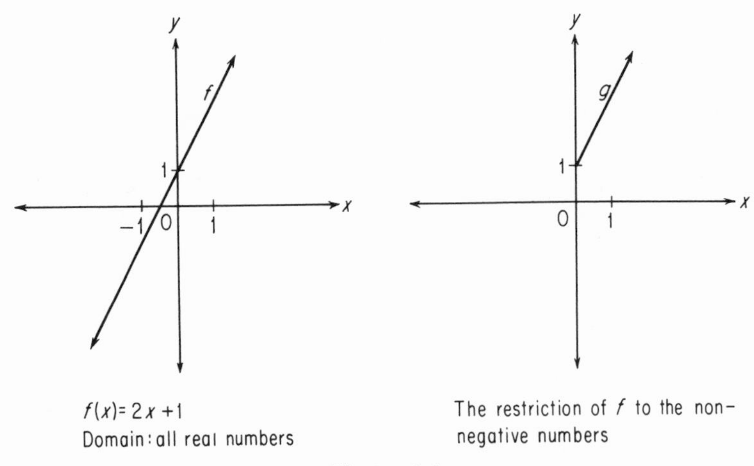

Figure 6-2

Definition: Suppose the domain of f is D and the domain of g is A, where A is a subset of D (Figure 6-3). If $f(x) = g(x)$ for every x in A, then g is called the **restriction of f to A**. Also, f is called an **extension of g to D.**

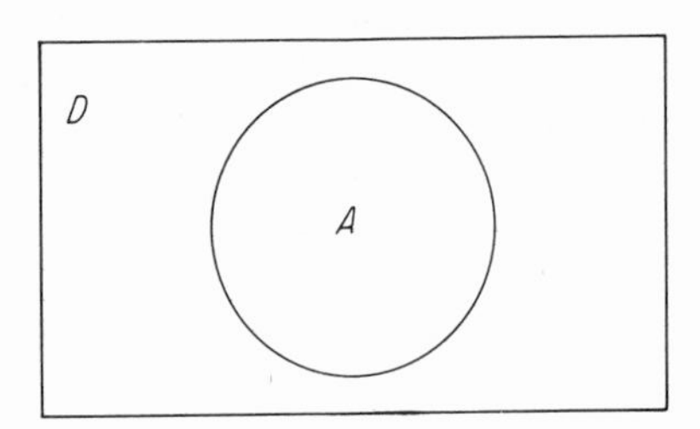

Figure 6-3

Note that if f is a given function with domain D and if A is a particular subset of D, then the restriction of f to A is a uniquely defined function. But if g is defined with domain A and if D is a set that contains A, there might be many different extensions of g to the domain D. One of the main topics of Chapter 7 is the most useful way to extend an exponential function from the domain of all *integers* to the domain of all real numbers.

6-2 EXERCISES

1. Let f be the function that has the set of all real numbers as its domain and has the rule $f(x) = 2 - 4x$. Sketch the graph of:

(a) The restriction of f to the non-negative numbers.
(b) The restriction of f to the integers.
(c) The restriction of f to the non-positive numbers.
(d) The restriction of f to $\{x: -2 \leqq x \leqq 5\}$.

2. Let f be the function that has the set of all real numbers as its domain and has the rule $f(x) = x^2$. Sketch the graph of:

(a) The restriction of f to the non-negative numbers.
(b) The restriction of f to the integers.
(c) The restriction of f to the negative numbers.
(d) The restriction of f to the domain $\{x: -2 \leqq x \leqq 5\}$.

3. The domain of a function has an important bearing on questions about possible maximum and minimum numbers in the range. For example, let $f(x) = 2x + 1$, domain: all real numbers. Let g be the restriction of f to the domain: $\{x: 1 \leqq x < 5\}$.

(a) Does the range of f have a maximum number? A minimum number?

(b) Does the range of g have a maximum number? A minimum number?

4. Let g be the function with domain the set of all *integers* and rule $g(x) = x$.

(a) Sketch the graph of g.

(b) Let $f(x) = [x]$, domain: all real numbers. (This is the "greatest integer function" of Exercise 4, Section 2-5.) Is f an extension of g to the set of all real numbers? Explain.

(c) Define two other extensions of g to the set of all real numbers, and sketch their graphs.

5. Let the domain of h be all non-negative numbers and let $h(x) = 2$.

(a) Sketch the graph of h.

(b) Define three different extensions of h to the set of all real numbers and sketch the graphs.

6. Let the domain of g be $\{0, 1, 2, 3\}$ and $g(n) = 2^n$.

(a) Sketch the graph of g.

(b) Define two different extensions of g to the interval $\{x: 0 \leqq x \leqq 3\}$ and sketch their graphs.

6-3 COMPOSITION OF FUNCTIONS

Suppose f and g are two functions such that the range of g is contained in the domain of f. Then a new function called the **composite function fg** can be defined as follows: The domain of fg is the same as the domain of g; the rule for fg is $fg(x) = f(g(x))$.

EXAMPLE 1: Let f and g each have the set of all real numbers as its domain, and let

$$f(x) = 2x + 1$$
$$g(x) = 3x - 2.$$

Then

$$fg(2) = f(g(2)) = f(4) = 9$$
$$fg(-3) = f(g(-3)) = f(-11) = -21.$$

For an arbitrary number x,

$$\begin{aligned} fg(x) &= f(3x - 2) \\ &= 2(3x - 2) + 1 \\ &= 6x - 3. \end{aligned}$$

The mappings of the functions f, g, and fg are shown in Figure 6-4.

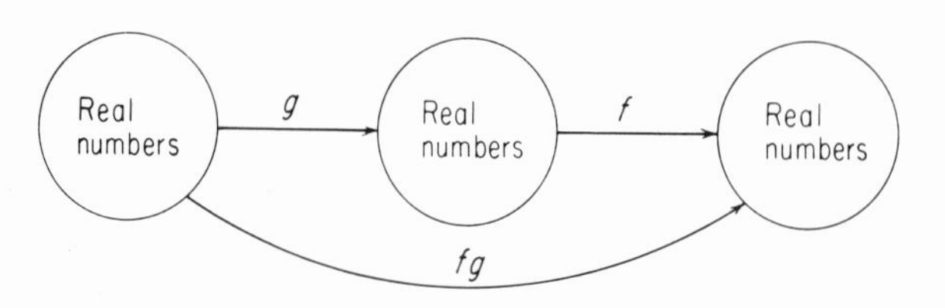

Figure 6-4

EXAMPLE 2: Suppose f and g denote two functions defined by

$f(x) = \sqrt{x}$, domain: the non-negative numbers

$g(x) = x^2$, domain: the real numbers.

Then fg is the function with domain the set of all real numbers and rule:

$$fg(x) = f(x^2) = \sqrt{x^2} = |x|$$

(See Figure 6-5.)

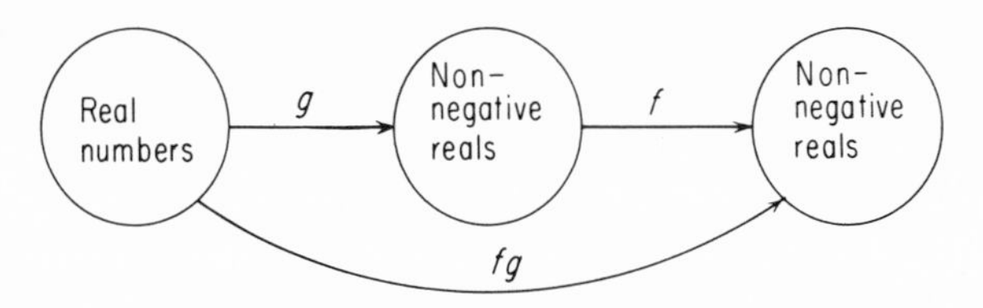

Figure 6-5

EXAMPLE 3: With f and g defined as in Example 2, gf is the function that has the set of all non-negative numbers as its domain and has the rule $gf(x) = g(\sqrt{x}) = (\sqrt{x})^2 = x$. Observe that the two composite functions gf and fg have different domains (Figure 6-6).

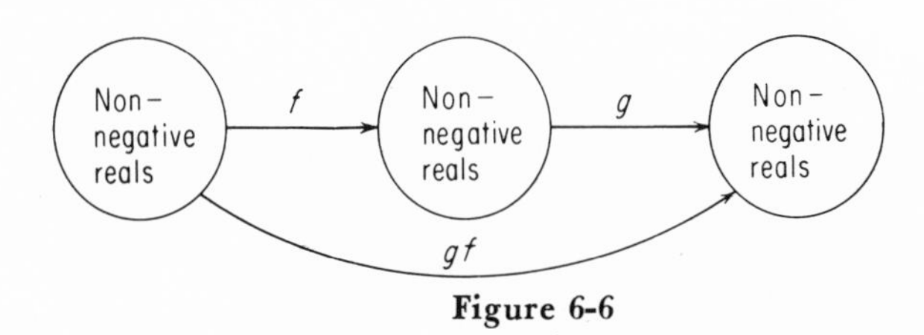

Figure 6-6

EXAMPLE 4: Let g have as its domain the set of all real numbers and rule $g(x) = x$. Let f have as its domain the set of all non-negative numbers and rule $f(x) = \sqrt{x}$. Then fg is not defined, since the range of g is not contained in the domain of f. (However, if the domain of g is restricted to a set of non-negative numbers, then the composite of f with such a restriction of g could be formed.)

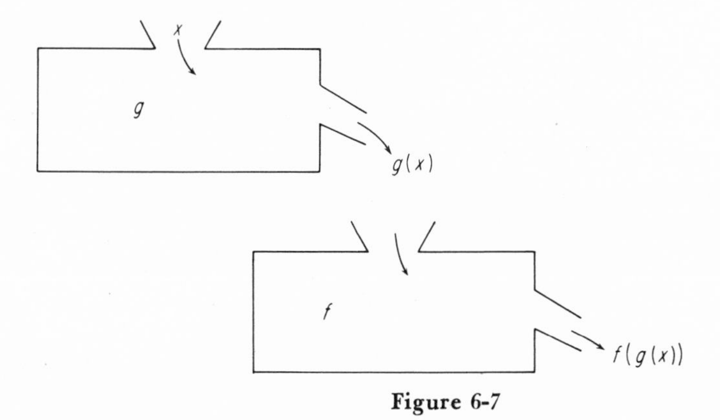

Figure 6-7

The "function machine" analogy of Chapter 1 helps to clarify composition of functions (Figure 6-7). Remember that for fg to be defined, the range of g must be contained in the domain of f.

The reader should be sure to work Exercises 9 and 13 in the list that follows. The general results obtained will be used in later chapters.

6-3 EXERCISES

1. Let

$$f(x) = 2 - x, \qquad \text{domain: all real numbers}$$

$$g(x) = 3x + 2, \qquad \text{domain: all real numbers.}$$

(a) $fg(2) = ?$
(b) $gf(2) = ?$
(c) Write a rule for fg and simplify it.
(d) Write a rule for gf and simplify it.

2. Let

$$f(x) = x^2, \qquad \text{domain: all real numbers}$$

$$g(x) = x + 3, \qquad \text{domain: all real numbers.}$$

(a) $fg(5) = ?$
(b) $gf(5) = ?$
(c) Write a rule for fg.
(d) Write a rule for gf.

3. Let

$$f(x) = |x|, \qquad \text{domain: all real numbers}$$

$$g(x) = |x - 2|, \qquad \text{domain: all real numbers.}$$

(a) $fg(-4) = ?$
(b) $gf(-4) = ?$
(c) Write the rule for fg and sketch the graph.
(d) Write the rule for gf and sketch the graph. [Hint: First sketch the graph of $h(x) = |x| - 2$.]

4. Give an example (different from Example 4) of two functions f and g such that fg is defined, but not gf.

5. Let f have as its domain the set of all real numbers and let $f(x) = 2x + 1$. Let g have the rule $g(x) = 1/(x - 2)$, domain: $\{x : x \neq 2\}$.
(a) What are the domain and rule for fg?
(b) $fg(5) = ?$

6. Let f and g have the set of all real numbers as their domain, and let $f(x) = 3x + 2$, $g(x) = \frac{1}{3}(x - 2)$.

(a) What are the domain and rule for fg?

(b) What are the domain and rule for gf?

7. The two functions of Exercise 6 have a special relationship to each other. Make up two more examples of pairs of functions that are similarly related.

8. (a) Suppose h is a function that has the set of all real numbers as its domain and has the rule $h(x) = \sqrt{x^2 + 1}$. Define two functions f and g such that $h = fg$.

(b) Suppose H is the function that has the set of all real numbers as its domain and has the rule $H(x) = (2x + 1)^5$. Define two functions F and G such that $H = FG$.

†**9.** Suppose that D is a set of numbers with the property that if x belongs to D then $-x$ also belongs to D. (For example, the set of all real numbers has this property.) Let g be a function with domain D and rule $g(x) = -x$. Let f be any function that has the same domain D. Then fg is defined and $fg(x) = f(-x)$.

(a) Show that the graph of fg is the reflection of the graph of f in the y-axis.

(b) Is the graph of fg congruent to the graph of f? Explain.

(c) Let the domain of f and g be $\{x\colon -2 \leqq x \leqq 2\}$ and let the rule for f be $f(x) = 3x + 1$. On the same set of axes sketch the graph of f and the graph of fg.

(d) Suppose that f is the function whose graph is shown here. Using just one set of axes, draw the graph of f, $-f$, fg, and $-fg$.

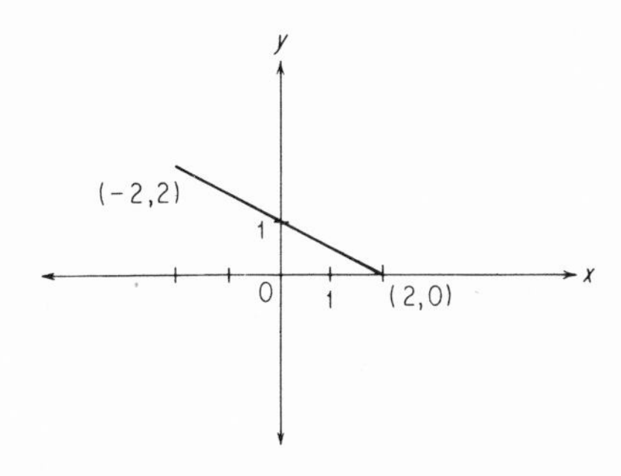

10. Let

$$g(x) = -x, \qquad \text{domain: all real numbers}$$

$$f(x) = x - 2, \qquad \text{domain: all real numbers.}$$

Using just one set of axes, sketch the graph of f, $-f$, fg, and $-fg$. (Suggestion: You may want to use pens of four different colors to distinguish the four graphs.)

11. Let

$$g(x) = x + 3, \qquad \text{domain: all real numbers}$$

$$f(x) = |x|, \qquad \text{domain: all real numbers.}$$

(a) What is the rule for fg?
(b) On the same set of axes sketch the graphs of f and fg.

12. Let

$$g(x) = x - 3, \qquad \text{domain: all real numbers}$$

$$f(x) = x^2, \qquad \text{domain: all real numbers.}$$

(a) What is the rule for fg?
(b) On the same set of axes sketch the graphs of f and fg.

13. Let $g(x) = x + c$, domain: some subset of the real numbers; let f be any real-valued function whose domain contains the range of g. Then fg is defined and $fg(x) = f(x + c)$. How is the graph of f related to the graph of fg? Are the two congruent? Explain.

6-4 INVERSE FUNCTIONS

Let f be a function with domain $\{a, b, c\}$ and rule given by the diagram in Figure 6-8. The range of f is the set $\{3, 5, 9\}$.

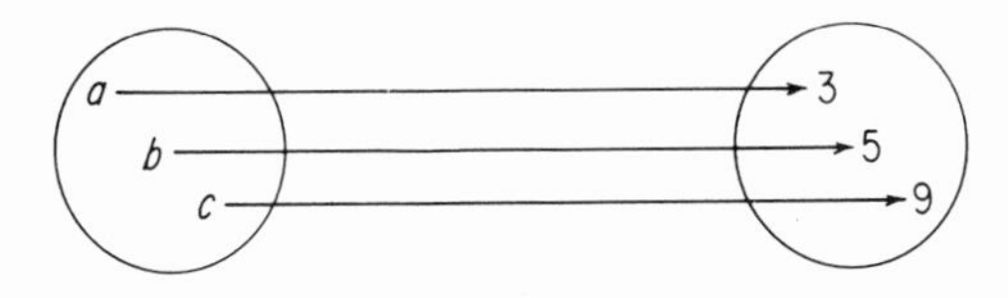

Figure 6-8

This simple function f, as shown in Figure 6-8, suggests another function, which we shall call g. Let the domain of g be the range of f, $\{3, 5, 9\}$; let the rule for g be given by the diagram in Figure 6-9. Thus g maps $f(a)$ onto a, $f(b)$ onto b, and $f(c)$ onto c.

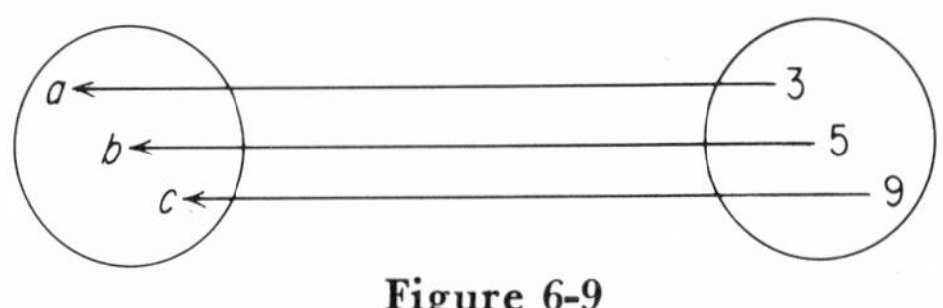

Figure 6-9

You will notice that not only is the domain of g taken to be the range of f, but also the range of g is the domain of f. Moreover, $fg(x) = x$ for every x in the domain of g, and $gf(x) = x$ for every x in the domain of f. These two functions are called **inverse functions**; each is the inverse of the other.

As a collection of ordered pairs,

$$f = \{(a, 3), (b, 5), (c, 9)\}$$

and

$$g = \{(3, a), (5, b), (9, c)\}.$$

Suppose we try to do the same thing with another function F, which has as its domain the set $\{a, b, c\}$ and has the rule given by the diagram shown in Figure 6-10. The range of F is the set $\{1, 3\}$.

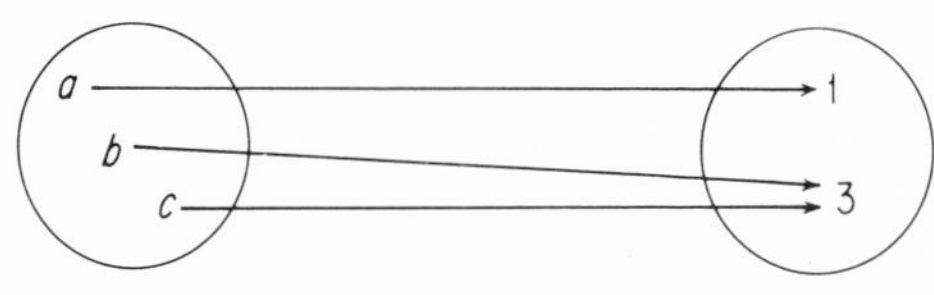

Figure 6-10

When we try to construct an inverse function G, we run into trouble: Should 3 be mapped onto b or onto c? No matter which choice is made, the range of G will be different from the domain of F. There is no way to construct G so that it will have all the properties that we want an "inverse function" to have.

If a function does not map any two different elements in its domain onto the same image, then an inverse function can be constructed. We will now give this property an official name, consider some more examples, and then give a formal definition of the inverse of such a function.

Definition: A function f with domain D is called a **one-to-one** function if and only if

$$\left.\begin{array}{r} x \text{ belongs to } D \\ \text{and } y \text{ belongs to } D \\ \text{and } x \neq y \end{array}\right\} \Rightarrow f(x) \neq f(y)$$

[Note: This is logically equivalent to saying that if $f(x) = f(y)$, then $x = y$.]

Every function (whether one-to-one or otherwise) has the property that each element in the *domain* has only *one image*. A one-to-one function has the additional property that each element in the *range* is the image of *only one element of the domain*.

A line parallel to the *y-axis* intersects the graph of a function (whether one-to-one or otherwise) in at most one point. A line parallel to the *x-axis* intersects the graph of a one-to-one function in at most one point. If a line parallel to the *x-axis* intersects the graph of f in two or more points, then f is *not* a one-to-one function.

EXAMPLE 1: Let f be the function that has the set of all real numbers as its domain and has the rule $f(x) = 2x + 1$. Then

$$f(x) = f(y) \Leftrightarrow 2x + 1 = 2y + 1 \Leftrightarrow x = y.$$

Thus, f is a one-to-one function. Similarly, it can be verified that *every linear* function is one-to-one. Observe that an arbitrary line parallel to the x-axis intersects the graph in only one point (Figure 6-11).

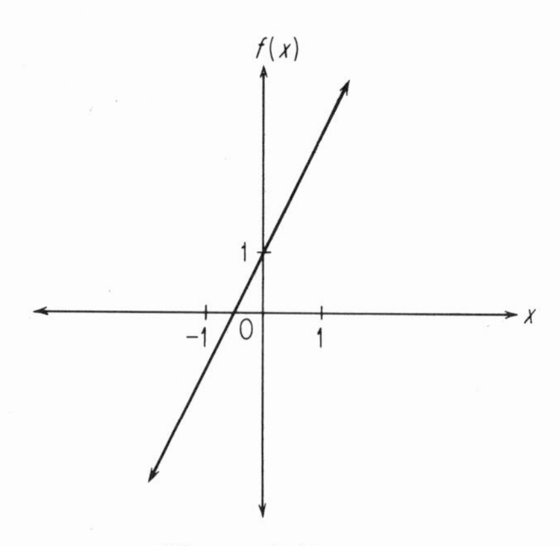

Figure 6-11

EXAMPLE 2: Let g be the quadratic function that has the set of all real numbers as its domain and has the rule $g(x) = x^2$. This function is *not* one-to-one, since $g(1) = g(-1)$, and, indeed, $g(x) = g(-x)$ for all x. Observe that some lines parallel to the x-axis intersect the graph in two points (Figure 6-12).

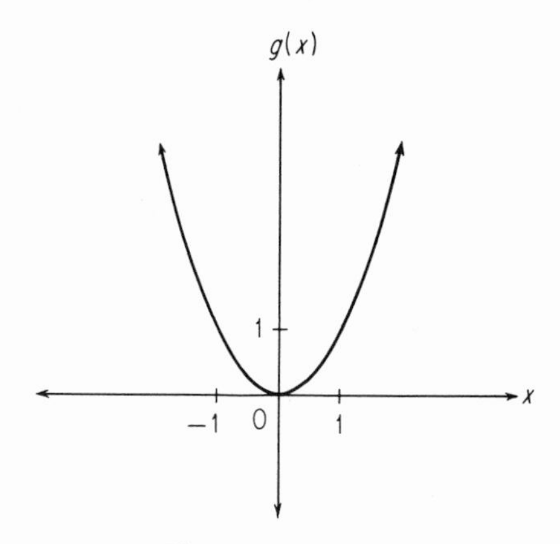

Figure 6-12

EXAMPLE 3: Let h be the restriction of g (Example 2) to the set of non-negative numbers (Figure 6-13). Then *h is* one-to-one. (Of course, there are other ways to restrict g so that the function thus formed will be one-to-one.)

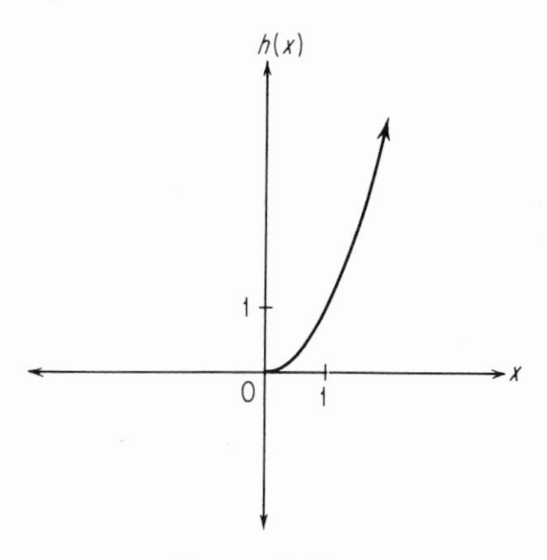

Figure 6-13

We are now ready to define the inverse of a one-to-one function.

Definition: Let f be a one-to-one function with domain D and range R. Construct another function g as follows: Let the domain of g be R; if y belongs to R, then $y = f(x)$ for some unique x in D. Define $g(y) = x$. The function g thus defined is called the **inverse** of f.

The student can now verify that:

(1) The range of g is the domain of f.

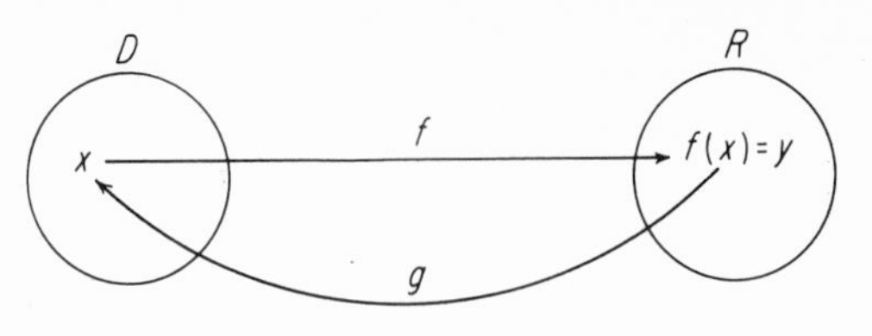

Figure 6-14

(2) The function g is a one-to-one function. The inverse of g is f.
(3) For every y belonging to R, $fg(y) = y$.

Once again the function machine analogy is helpful. Suppose f is a one-to-one function and g its inverse. Machine f accepts x and produces $f(x)$; machine g accepts $f(x)$ and produces x—the very thing that was fed into f to begin with (Figure 6-15). Moreover, if the order of the two machines is reversed, a similar result is obtained (Figure 6-16).

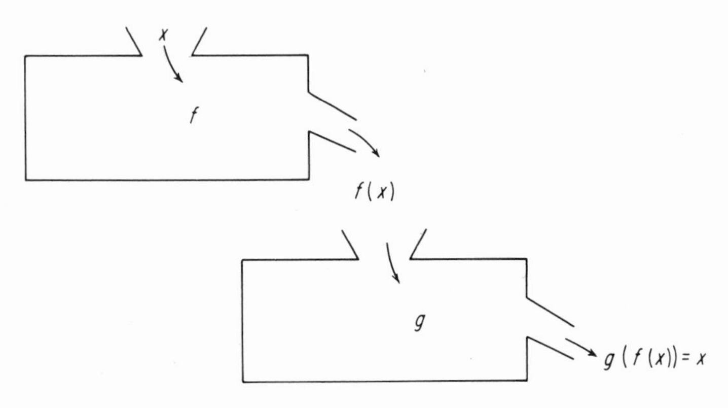

Figure 6-15

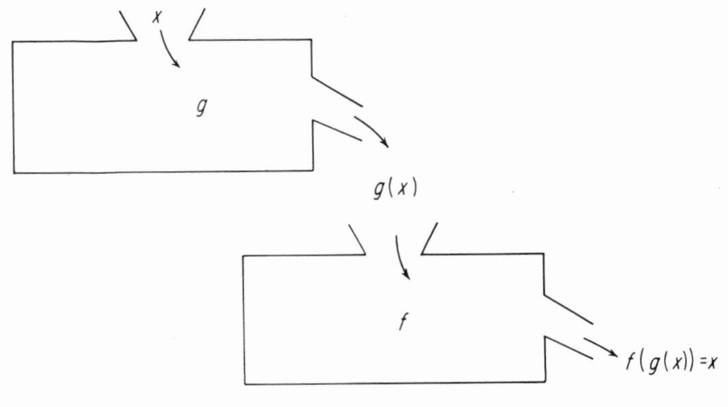

Figure 6-16

EXAMPLE 4: Let f be the function that has the set of all real numbers as its domain and has the rule $f(x) = 2x + 1$. We know this linear function is one-to-one; therefore, its inverse function g can be constructed. The domain of g is all real numbers. Observe that $2x + 1 = y \Leftrightarrow x = \frac{1}{2}(y - 1)$. So the rule for g is given by $g(x) = \frac{1}{2}(x - 1)$. (See Figure 6-17.)

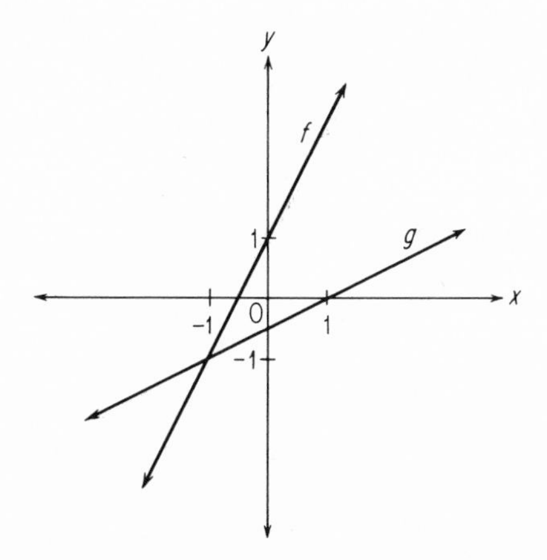

Figure 6-17

EXAMPLE 5: Let f be the function with domain $\{x: 0 \leqq x \leqq 3\}$ and rule $f(x) = 3x$. We know that f is a one-to-one function with range $\{y: 0 \leqq y \leqq 9\}$. The inverse function g has domain $\{x: 0 \leqq x \leqq 9\}$ and rule $g(x) = \frac{1}{3}x$. (See Figure 6-18.)

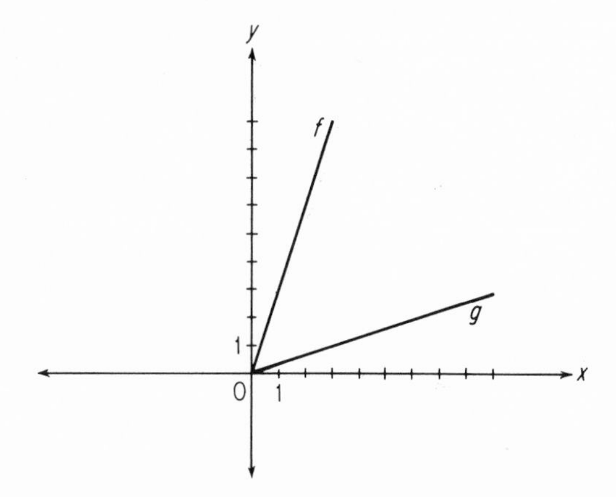

Figure 6-18

EXAMPLE 6: Let f be the function with domain $\{x: 0 \leqq x \leqq 2\}$ and rule $f(x) = x^2$. The range of this function is $\{y: 0 \leqq y \leqq 4\}$. The inverse function g has domain $\{x: 0 \leqq x \leqq 4\}$ and has the rule $g(x) = \sqrt{x}$. (We know this because f maps $\sqrt{x}$ onto x; so g maps x onto $\sqrt{x}$. See Figure 6-19.)

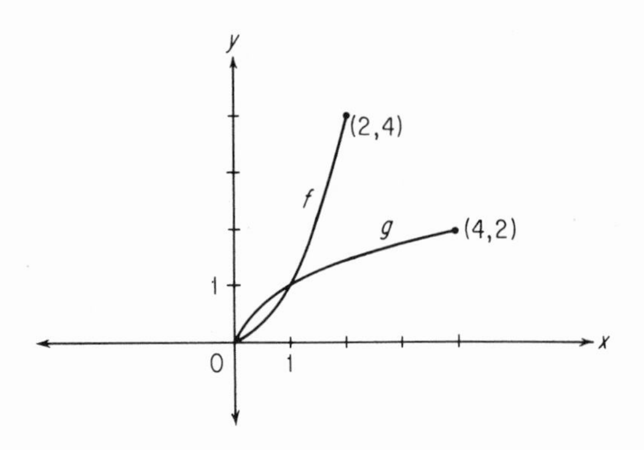

Figure 6-19

Suppose f and g denote inverse functions. A point (x, y) is on the graph of f if and only if (y, x) is on the graph of g. It has been observed in Section 4-1 that the point (b, a) is the reflection of the point (a, b) in the line $\{(x, y): y = x\}$. This leads to a helpful observation about the graph of a one-to-one function and the graph of its inverse: Each is the reflection of the other in the line $\{(x, y): y = x\}$. Intuitively, it is also clear that the two graphs are congruent (Figure 6-21). In Exercise 8 the student is asked to describe a rigid motion that will take the graph of f onto that of its inverse.

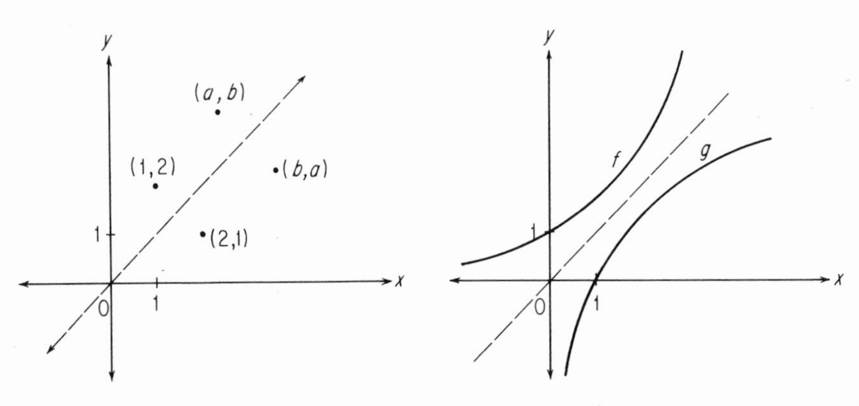

Figure 6-20 **Figure 6-21**

6-4 EXERCISES

1. Let $f(x) = |x - 2| + 1$, domain: all real numbers.
(a) Show that this is not a one-to-one function by giving two numbers that have the same image.
(b) Suggest a restriction of the domain so that the function with the same rule but restricted domain will be one-to-one.

2. Select the functions that are one-to-one.

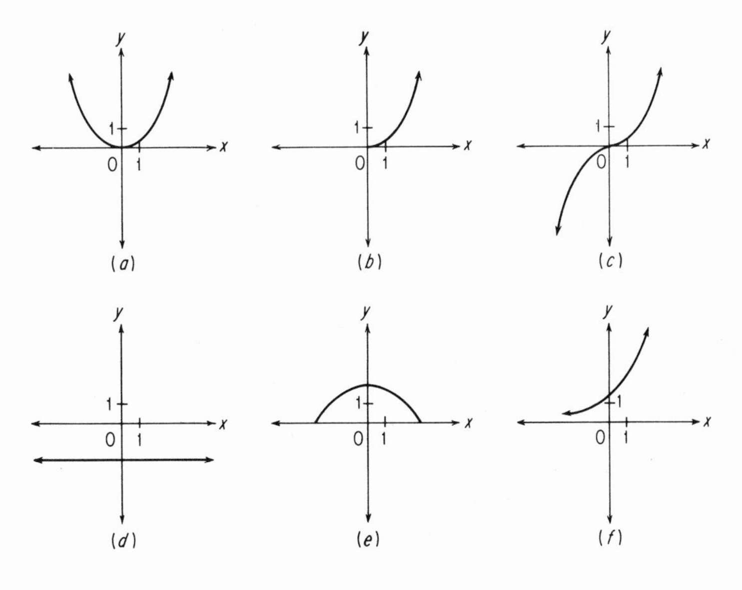

3. Let $A = \{1, 2, 3, 4\}$ and $B = \{5, 6, 7, 8\}$. Let f be a subset of $A \times B$ defined as follows:

$$f = \{(1, 5), (2, 6), (3, 8), (4, 7)\}$$

List the elements (ordered pairs) of the function that is the inverse of the function f.

4. Suppose A and B are defined as in Exercise 3. Suppose

$$g = \{(1, 6), (2, 5), (3, 7), (4, 6)\}$$

Explain why the function g does not have an inverse function.

5. Let $f(x) = \frac{1}{2}x + 3$, domain: all real numbers. This function is one-to-one. Let g denote its inverse.
(a) $g(7) = ?$ $g(3) = ?$
(b) What is the rule for g?

6. Each of these functions is one-to-one. Construct the inverse function, specifying both the domain and the rule. For each, sketch the graph of the function and of its inverse.

(a) f has as its domain the set of all real numbers and has the rule $f(x) = 4x + 1$.
(b) g has as its domain $\{x: -2 \leqq x \leqq 4\}$ and $g(x) = 2 - 5x$.
(c) h has as its domain the set of all real numbers and $h(x) = x^3$.
(d) F has as its domain the set of all positive numbers and has the rule $F(x) = 1/x$.
(e) G has as its domain the set of all non-negative numbers and has the rule $G(x) = \sqrt{x}$.

7. Sketch the graph of the inverse of each of these three one-to-one functions.

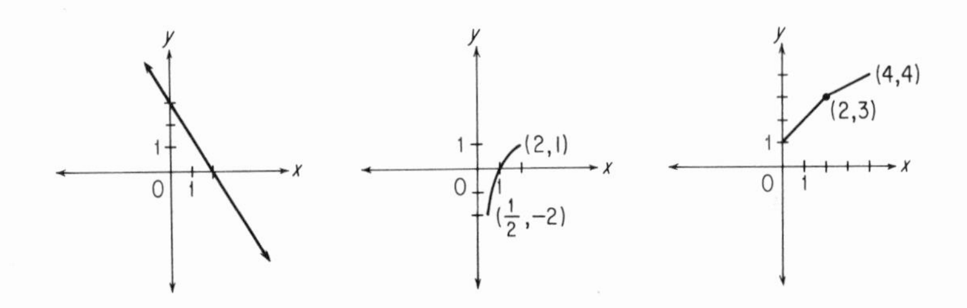

8. Suppose that f and g are inverse functions. Describe a rigid motion of the plane that will take the graph of f onto the graph of g.

9. Suppose f is the function with domain $\{-3, -2, -1, 0, 1, 2, 3\}$ and rule $f(n) = 2^n$. This function is a one-to-one function.

(a) Define the inverse function by giving a table of values.
(b) Sketch the graph of the function and of its inverse.

10. Suppose f is the function with domain $\{-4, -3, -2, -1, 0, 1, 2, 3, 4\}$ and rule $f(n) = 10^n$.

(a) Make a table of values of f.

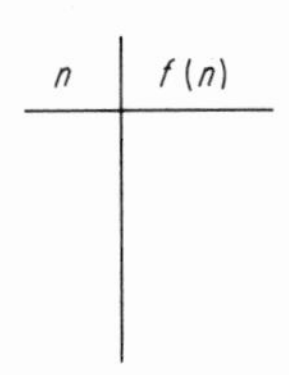

(b) Let g denote the inverse of f. What is the domain of g? Make a table of values.

11. Here are the graphs of four one-to-one functions. Reflect each graph in the line $\{(x, y): y = x\}$ to obtain the graph of the inverse function.

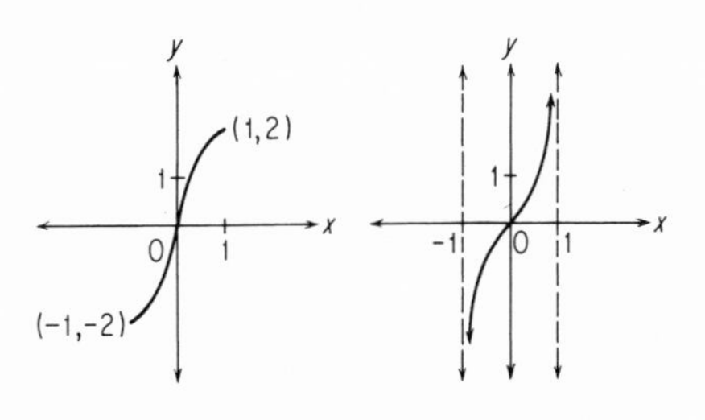

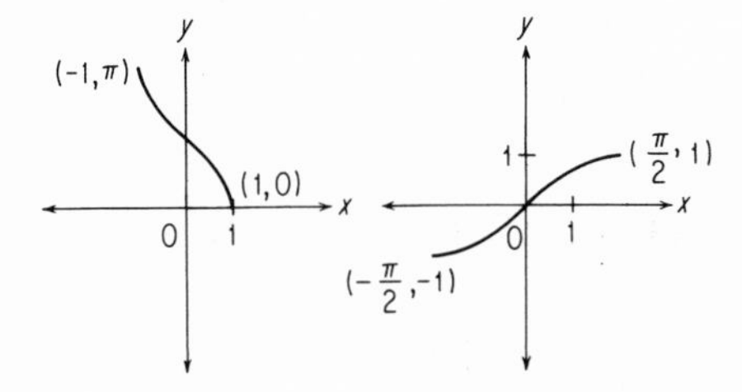

7

Exponential Functions

7-1 REVIEW OF INTEGRAL EXPONENTS

It is assumed that the student has some knowledge of integral exponents. We present a brief summary of the most important definitions and theorems.

Definitions: **1.** Let a stand for any number and let n denote a positive integer. Then $a^n = a \cdot a \cdot a \cdot a \cdot \cdots \cdot a$ (to n factors).

2. If $a \neq 0$, then $a^0 = 1$.

3. If $a \neq 0$, and n is a positive integer, then $a^{-n} = 1/a^n$.

Theorem 7-1: If $a \neq 0$ and m and n denote integers, then

1. $a^m \cdot a^n = a^{m+n}$.

2. $a^m/a^n = a^{m-n}$.

3. $a^{mn} = (a^m)^n$.

The three properties of exponents in Theorem 7-1 will be important for nearly everything that follows. In addition, there are two other theorems that the student should know.

Theorem 7-2: If $ab \neq 0$ and n denotes an integer, then

1. $(ab)^n = a^n \cdot b^n$.

2. $(a/b)^n = a^n/b^n$.

Theorem 7-3: If $a \neq 1$ and $a \neq -1$ and $a \neq 0$ and if $a^m = a^n$, where m and n are integers, then $m = n$.

It is suggested that the reader convince himself of the validity of these three theorems.

7-1 EXERCISES

1. Write each of these numbers in the form 10^n, where n is an integer:

(a) One million (b) One billion
(c) Ten billion (d) 0.01
(e) 0.0001 (f) 100^3
(g) 10000^5 (h) $1000 \times 10{,}000$
(i) $(0.01)^3$ (j) $2^4 \cdot 5^4$
(k) $4^8 \cdot 5^{16}$ (l) $100^6/1000^8$
(m) $(0.01)^8/100$ (n) 1
(o) 100^{-2} (p) $(0.01)^{-4}$
(q) $(100^{-2})^{-4}$ (r) $5^{-2} \cdot 2^{-2}$

2. Write each of these numbers in the form 2^n, where n is an integer:

(a) 16 (b) 256
(c) 1024 (d) 4^3
(e) 8^7 (f) $\frac{1}{2}$
(g) $\frac{1}{4}$ (h) $(\frac{1}{4})^3$
(i) $(\frac{1}{32})^{-4}$ (j) $(\frac{1}{4})^{-3}$
(k) 1 (l) $(\frac{1}{8})^{-2}$

3. By definition, $a^0 = 1$ if $a \neq 0$. Give a reason why this definition is desirable.

4. Write without the use of exponents and simplify as much as possible.

Example: $3.4 \times 10^5 \times 10^{-3} = 3.4 \times 10^2 = 340.$

(a) $10^9 \times 10^{-4}$ (b) $10^{-5}/100^{-2}$
(c) 1.45×10^{-8} (d) $(100^8 \times 1000^{15})/10^{60}$
(e) $[(1876 \times 678)^0 \times 2^{50}]/4^{20}$ (f) $7^{20} \times 14^{-19} \times 2^{21}$

5. The numbers in this list have been selected so that they can be written in the form $2^n \cdot 5^m$, where m and n are integers. Write them in this form.

(a) 20 (b) $\frac{1}{20}$
(c) 100 (d) 100^{15}
(e) $(0.01)^{15}$ (f) 80
(g) $\frac{1}{80}$ (h) $100{,}000^8$

6. Solve these equations. Observe that Theorem 7-3 assures us that each equation has at most one solution.

(a) $2^n = 1024$
(b) $4^n = 1024$
(c) $2^n = \frac{1}{8}$
(d) $3^n = 1$
(e) $3^n = 0$
(f) $2^{-n} = 1024$
(g) $10^n = 1{,}000{,}000{,}000$
(h) $10^n = 0.0001$
(i) $100^n = 1{,}000{,}000$
(j) $2 \cdot 3^n = 18$
(k) $5 \cdot 2^n = \frac{5}{2}$
(l) $(10^6)^n = 10^{48}$
(m) $100^n = 1000^4$
(n) $8^n = 4^{12}$
(o) $2^n = -1$
(p) $3^n = 9^{20}$

7. There is no integer n such that $4^n = 8$. Later in this chapter we shall consider an exponential function with domain the set of all real numbers and rule $f(x) = 4^x$. In this extended domain there is a number x such that $4^x = 8$. In general, if $a > 0$ and x and y denote arbitrary real numbers, there is a definition of a^x that satisfies the condition $a^{xy} = (a^x)^y$. Assume all this and assume that each of the following equations has a unique solution. Solve.

(a) $4^x = 8$ [Hint: $4^x = 8 \Leftrightarrow (2^2)^x = 2^3$]
(b) $100^x = 1000$
(c) $16^x = 4$
(d) $1000^t = 10$
(e) $9^s = 27$
(f) $64^y = 32$
(g) $1000^x = 0.1$
(h) $100^y = 0.001$
(i) $9^x = 243$
(j) $32^x = 16$
(k) $100^x = 10$
(l) $1000^x = 10$

7-2 SCIENTIFIC NOTATION

Our numeral system is a decimal system; numbers that can be written in the form 10^n (where n is an integer) have a special role to play. Here are a few of these numbers, each one written in two forms:

$$10^{-4} = 0.0001$$
$$10^{-3} = 0.001$$
$$10^{-2} = 0.01$$
$$10^{-1} = 0.1$$
$$10^{0} = 1$$
$$10^{1} = 10$$
$$10^{2} = 100$$
$$10^{3} = 1000$$
$$10^{4} = 10000.$$

The numbers that can be written in the form 10^n, where n is an integer, are only sparsely distributed among the positive real numbers. Many numbers cannot be written as integral powers of ten. The number 7682, for example, is between 10^3 and 10^4; we cannot write it in the form 10^n, where n denotes an integer. But we *can* write 7682 in the form $k \times 10^n$, where $1 \leqq k < 10$:

$$7682 = 7.682 \times 10^3.$$

Similarly,

$$93{,}000{,}000 = 9.3 \times 10^7$$

$$0.00000176 = 1.76 \times 10^{-6}$$

$$0.0000000024 = 2.4 \times 10^{-9}.$$

Definition: A positive number is said to be written in **scientific notation** if and only if it is written in the form $k \times 10^n$, where n is an integer and $1 \leqq k < 10$.

There are several reasons why it is sometimes best to write numbers in scientific notation. Perhaps the most obvious reason is that scientific notation provides a simple way to write a number that is quite large or quite small. For example, it is simpler to write Avogadro's number in the form

$$6.02 \times 10^{23}$$

than it is to write it in the form

$$602000000000000000000000.$$

There is another reason for employing scientific notation. A measurement of physical quantity such as length, weight, or time is at best an approximation to the "true" value of the quantity. Scientific notation provides one convenient way for a scientist to report a measurement so as to indicate the magnitude of the possible error of the measurement. Consider the two statements:

(1) The distance between A and B is 47,000,000 miles.
(2) The distance between A and B is 4.700×10^7 miles.

The second statement contains more information than the first, even though the numbers 47,000,000 and 4.700×10^7 are identical. The reader of statement (1) feels fairly sure that the vast distance 47,000,000 miles has not been measured correct to the nearest mile, but he is left to wonder just what the maximum possible error is. On the other hand the (knowledgeable) reader of statement (2) understands that when a measurement is reported as 4.700×10^7 miles, this means that all the digits in 4.700 are believed to be correct; the true measurement is between 4.7005×10^7 miles and 4.6995×10^7 miles.

If an author writes Avogadro's number as

$$6.02 \times 10^{23}$$

he is telling the reader that there is good reason to believe that all the digits in 6.02 are correct; the true value of Avogadro's number is between

$$6.025 \times 10^{23}$$

and

$$6.015 \times 10^{23}.$$

Therefore, the difference between 6.02×10^{23} and the true Avogadro's number does not exceed 0.005×10^{23} in absolute value. The number 0.005×10^{23} is known as the **absolute error** of this measurement.

If the speed of light is written as

$$3.0 \times 10^{10} \text{ cm/sec}$$

this implies that the true speed of light is between

$$3.05 \times 10^{10} \text{ cm/sec}$$

and

$$2.95 \times 10^{10} \text{ cm/sec}.$$

Thus the absolute error of this measurement is

$$0.05 \times 10^{10} \text{ cm/sec}.$$

If a measurement is written in scientific notation, $k \times 10^n$, where $1 \leqq k < 10$ (with k written as a decimal having a finite number of digits), the digits in k are called **significant digits.** The number of digits in k is also the number of significant digits of the measurement.

Here are several measurements. For each the number of significant digits has been listed, and also the absolute error of the measurement.

Measurement	*Number of Significant Digits*	*Absolute Error*
2.71×10^8 miles	3	500,000 miles
2.710×10^8 miles	4	50,000 miles
2.710000×10^8 miles	7	50 miles
2.71×10^{-4} cm	3	0.0000005 cm
2.710×10^{-4} cm	4	0.00000005 cm

Scientific notation is useful, but the reader should not infer that *every* measurement must be reported in scientific notation in order to give a clear indication of what the absolute error is. For example, if the measurement of a certain distance is reported as

7200 miles (to the nearest hundred miles)

this implies that the true distance is between

7150 miles

and

7250 miles.

So the absolute error of the measurement is 50 miles.

If a length is reported as

72.1 cm (to the nearest tenth of a centimeter)

this implies that the true length is between

72.15 cm

and

72.05 cm.

The absolute error of the measurement is 0.05 cm.

If the length is reported as

$$72.1 \text{ cm} \pm 0.005 \text{ cm}$$

this implies that the true length is between

72.105 cm

and

72.095 cm.

The absolute error is 0.005 cm.

7-2 EXERCISES

1. Write each of these numbers in scientific notation.

(a) 43,200.
(b) One million.
(c) 0.1390.
(d) 35.018.
(e) 0.003576.
(f) 0.000000084.
(g) $10^{15} + 10^{16}$ [Hint: $10^{15} + 10^{16} = 10^{16}(10^{-1} + 1)$].
(h) $3 \times 10^{12} + 5 \times 10^{14}$.

(i) $(2 \times 10^{16})(11 \times 10^{7})$.
(j) $10^{21} + 10^{24}$.
(k) $3.1 \times 10^{7} + 4.2 \times 10^{6}$.
(l) $5.61 \times 10^{8} - 2.30 \times 10^{7}$.

2. Give a convincing argument that:
(a) Every positive number can be written in scientific notation—that is, in the form $k \times 10^n$, where n is an integer and $1 \leqq k < 10$.
(b) If a given positive number is written in scientific notation, there is just one correct choice of the integer n and just one correct choice of the number k.

3. Write each of these numbers in the form $k \times 2^n$, where n is an integer and $1 \leqq k < 2$.

Example: $12 = \frac{12}{8} \times 8 = 1.5 \times 2^3$.

(a) 10 (b) 42
(c) 50 (d) 100
(e) 200 (f) 0.8

4. Write each of these measurements in scientific notation. How many significant digits are there in each?
(a) 93,000,000 miles (given an absolute error of 500,000 miles).
(b) 93,000,000 miles (to the nearest 1000 miles).
(c) 245,000 lb $\pm$ 500 lb.
(d) 245,000 lb (to the nearest 10 lb).
(e) (750 ± 5) lb.
(f) (0.0089 ± 0.00005) cm.
(g) 0.00890 cm (given an absolute error of 0.000005 cm).
(h) 0.19 cm (to the nearest hundredth of a centimeter).

7-3 SOME APPLICATIONS OF EXPONENTIAL FUNCTIONS

The definitions of Section 7-1 assign meaning to a^n for every integer n $(a \neq 0)$. This allows us to talk about a function f, for example, whose rule is $f(n) = 2^n$, or a function g whose rule is $g(n) = 6(\frac{1}{2})^n$ *provided the domain is restricted to some subset of the integers*. Such functions have applications in a wide variety of studies—from physics to finance and from biology to economics.

EXAMPLE 1: Suppose \$100 is invested at 5% compounded annually. How much is this investment worth after 10 years?

Solution: In analyzing this question it is natural to construct a rule for $A(n)$—the amount of the investment after n years, where n denotes an arbitrary positive integer.

$$A(1) = 100 + 0.05(100) = 100(1.05)$$
$$A(2) = 100(1.05) + 0.05[100(1.05)] = 100(1.05)^2$$
$$A(3) = 100(1.05)^2 + 0.05[100(1.05)^2] = 100(1.05)^3$$
$$\ldots$$
$$A(n) = 100(1.05)^n$$

In particular, $A(10) = 100(1.6289)$, where the number in parentheses is obtained from the table preceding the exercises of this section. So we conclude that after 10 years the amount of the investment is \$162.89. The function rule developed here is a typical exponential rule; it has the form $f(n) = k \cdot a^n$, where a is a positive number different from 1.

EXAMPLE 2: The 1950 census showed the population of a certain city to be 10,000 people. Since that time the population has increased approximately 8% every 5 years. Assuming that this growth pattern is maintained, predict the population in 1980.

Solution: The required function rule is $P(n) = 10{,}000(1.08)^n$, where n denotes the number of five-year periods that have elapsed since 1950. Therefore, the approximate population for 1980 should be $P(6) = 10{,}000(1.08)^6$. Using the table preceding the exercises of this section (and rounding off to the nearest thousand) we predict the 1980 population of the city to be about 16,000.

EXAMPLE 3: Assume that the city of Example 2 had been experiencing 8% growth each 5 years for several decades prior to 1950. Approximately what was the population in 1925?

Solution: The function rule $P(n) = 10{,}000(1.08)^n$ is valid when n is a negative integer. The approximate population in 1925 must have been $P(-5) = 10{,}000(1.08)^{-5}$. This is approximately 7000.

EXAMPLE 4: The half-life of radioactive bismuth is approximately 5 days. This means that at the end of 5 days one-half a given quantity of bismuth will have disintegrated, leaving only one-half of the original quantity (by weight). If a sample of radioactive bismuth weighs k grams, what will be its weight after 30 days?

Solution: The function rule is $W(n) = k \cdot (\frac{1}{2})^n$, where n denotes the number of five-day periods that elapse. So after 30 days the weight of the sample will be $W(6) = k \cdot (\frac{1}{2})^6$. Thus $\frac{1}{64}$ of the sample will remain.

This table of approximate values is useful for some of the examples and exercises of this section. More extensive tables of a similar nature are found in many handbooks.

n	$(1.05)^n$	$(1.08)^n$
−5	0.7835	0.6806
−4	0.8227	0.7350
−3	0.8638	0.7938
−2	0.9070	0.8573
−1	0.9524	0.9259
0	1.0000	1.0000
1	1.0500	1.0800
2	1.1025	1.1664
3	1.1576	1.2597
4	1.2155	1.3605
5	1.2763	1.4693
6	1.3401	1.5869
7	1.4071	1.7138
8	1.4775	1.8509
9	1.5513	1.9990
10	1.6289	2.1589

7-3 EXERCISES

1. Let f have domain $\{-3, -2, -1, 0, 1, 2, 3\}$ and rule $f(n) = 2^n$.
(a) Sketch the graph of f and of $-f$.
(b) $f(2) \cdot f(1) = ?$
(c) $f(2) \cdot f(0) = ?$
(d) $2f(-1) = ?$
(e) $[f(2)]^2 = ?$
(f) $\dfrac{f(2)}{f(1)} = ?$

(g) $\dfrac{f(2)}{f(-1)} = ?$

2. Let g have domain $\{-3, -2, -1, 0, 1, 2, 3\}$ and rule $g(n) = (\frac{1}{2})^n$.

(a) Sketch the graph of g and of $-g$.

(b) What is the relation between g and the function called f in Exercise 1?

3. \$1000 is invested at 8% compounded annually. What will it be worth:

(a) After 5 years? (b) After 10 years?

(c) After n years?

4. It is desired to invest enough money right now at 5% compounded annually so that its value 5 years from now will be \$10,000. How much should be invested?

5. How much should be invested now at 8% compounded annually so that it will be worth \$100,000 in 5 years?

6. Mr. A is offered two jobs. The first offer is \$25 for the first day's work, with an increase thereafter of \$20 each day. The second job offers only \$5 for the first day; but the pay doubles each day. Which job should Mr. A choose? It does not take long to decide that the second offer is the better if the job lasts long enough.

(a) Write the rule for the function that will express Mr. A's daily pay (in dollars) if he accepts the first job offer. Then write the corresponding rule for the second job offer.

(b) What is the number of the day on which the second job would pay better than the first?

7. A rubber ball is dropped from a height of 10 feet. On the first bounce it reaches a height of 5 feet. Each time thereafter it bounces to a height of one-half that reached on the previous bounce.

(a) What height is reached on the fifth bounce?

(b) What height is reached on bounce number n?

8. A certain radioactive substance decomposes in such a way that 90% of the material is left after one year. If k milligrams of the material are present at a given time, write the rule for the function that tells how much will be present t years later.

9. A biologist has reason to believe that the number of bacteria in a certain colony will have an exponential growth. Let $N(t)$ denote the number of bacteria in the colony at time t, where t is measured in hours. If the growth is exponential, this means there are positive numbers k and a such that $N(t) = k \cdot a^t$. Suppose the biologist knows that $N(0) = 10{,}000$ and $N(2) = 22{,}500$. What should he predict for the number of bacteria in the colony at the end of 4 hours?

7-4 THE RATIONAL NUMBERS AS A SUBSET OF THE REAL NUMBER SYSTEM

At this point we digress from our development of exponential functions to review some basic properties of the real number system—properties that are important for the reader who wants to understand the later sections of this chapter. We shall not prove many of the facts that are stated in this and later sections. The reader can understand the definitions and, by working with specific examples, can understand what facts are asserted to be true even though the proof of these facts is not presented. For a more detailed development of the properties of the real number system, see references [6] and [7] in the bibliography.

Every real number can be written in a decimal form. Some numbers have a decimal form that has a last non-zero digit and is called a **terminating decimal.** For example:

$$2^8 = 256$$
$$\tfrac{1}{2} = 0.5$$
$$\tfrac{4}{5} = 0.8$$
$$\tfrac{1}{25} = 0.04$$
$$\tfrac{9}{5} = 1.8$$

Other numbers have a decimal form that has no last non-zero digit.

$$\tfrac{1}{3} = 0.33333\ldots$$
$$\tfrac{1}{15} = 0.066666\ldots$$
$$\sqrt{2} = 1.4142\ldots$$
$$\pi = 3.14159\ldots$$

Three dots placed at the right of a decimal signify that there is no last digit: The decimal is called an **infinite decimal**.

For the numbers $\frac{1}{3}$ and $\frac{1}{15}$ the decimal form can be obtained by the process of division. For the numbers $\sqrt{2}$ and π the decimal form is more difficult to obtain. The main point of this discussion is that every real number does *have* a decimal form and that this decimal form is either terminating or is infinite.

Unfortunately, some numbers have more than one decimal form. For example, 3 and 2.99999. . . are two different decimals that denote the same number. We would like to rule out of our discussion any infinite decimal, such as 2.99999. . . , that has only "nines" beyond a certain digit and any infinite decimal, such as 3.0000. . . , that has only "zeros" beyond a certain digit. In order to do this we shall make the following definition: By the **decimal expansion** of a number we shall mean a decimal form of the number that is not an infinite decimal:

(1) With only a finite number of its digits different from 9.
(2) With only a finite number of its digits different from 0.

For example,

3	2.45
3.99	1.555. . .
0.131131113. . .	0.1656565. . .
1.414. . .	3.1415. . .

are decimal expansions of numbers. But we shall not call 1.9999. . . a decimal expansion of 2, and we shall not call 0.159999. . . a decimal expansion of 0.16.

With this agreement about the meaning of the decimal expansion of a number, the following significant statement is valid: Every real number has a *unique* decimal expansion. Furthermore, every decimal, whether terminating or infinite, denotes a real number.

Some infinite decimals have a repeating pattern; others, such as the decimal expansions of $\sqrt{2}$ and π, are not repeating decimals.

The infinite decimal expansion of $\frac{1}{3}$ has a repeating pattern—every digit that follows the decimal point is 3:

$$\tfrac{1}{3} = 0.333333\ldots$$

The repeating cycle in an infinite repeating decimal need not begin with the first digit following the decimal point. For example:

$$\tfrac{9}{55} = 0.1636363\ldots$$

To remove any doubt in the mind of a reader as to whether an infinite decimal as written is intended to be a repeating decimal, we shall adopt the convention of placing a bar over the digits of a repeating cycle the first time the cycle occurs in the decimal expansion. If a bar is placed over a digit or over several consecutive digits, the reader is to interpret the decimal as a repeating decimal. Thus each of the following numerals is a repeating decimal:

$$0.\overline{3}\ldots = 0.3333\ldots$$
$$0.1\overline{63}\ldots = 0.1636363\ldots$$
$$2.71\overline{8}\ldots = 2.718888\ldots$$
$$0.12\overline{563}\ldots = 0.12563563563\ldots$$

In Section 1-2 a rational number was defined as a real number that can be written as the quotient of two integers. The following theorem asserts that the set of all numbers that can be represented by terminating decimals or by infinite repeating decimals is the set of rational numbers.

Theorem 7-4: If a real number can be written as the quotient of two integers, then the decimal expansion of the number is a terminating decimal or an infinite repeating decimal. Conversely, every terminating decimal and every infinite repeating decimal can be written as the quotient of two integers.

In the light of this theorem we can test a real number to see whether it is a rational number in either one of two ways:

(1) Can the number be expressed as the quotient of two integers?

(2) Is the decimal expansion of the number a terminating decimal or an infinite repeating decimal?

A number that meets either of these two tests is a rational number.

The set of rational numbers includes every number that a non-scientist is likely to encounter in the pursuit of everyday affairs. Included in the set are:

(1) Every integer.

(2) Every common fraction a/b in which both a and b are integers with $b \neq 0$.

(3) Every terminating decimal.

(4) Every infinite repeating decimal.

It should be remarked also that the rational numbers form an important algebraic system (called a **number field**) that is closed with respect to each of the four fundamental operations of arithmetic. This means:

(1) The sum of two rational numbers is a rational number.

(2) The difference of two rational numbers is a rational number.

(3) The product of two rational numbers is a rational number.

(4) The quotient of two rational numbers (where the divisor is not zero) is a rational number.

7-4 EXERCISES

1. Write the decimal expansion for each of these rational numbers:

(a) $\frac{1}{6}$ (b) $\frac{13}{12}$
(c) $\frac{7}{8}$ (d) $\frac{1}{9}$
(e) $\frac{1}{7}$ (f) $\frac{2}{15}$
(g) $\frac{7}{6}$ (h) $\frac{27}{20}$

2. Let a and b denote integers with $b > 0$. Give a convincing argument that the rational number a/b has a decimal expansion that is a terminating decimal or an infinite repeating decimal.

3. Write each of these decimals as a number in the form a/b, where a and b are integers:

(a) 3.45 (b) 0.0067
(c) $0.\overline{6}\ldots$ (d) $0.\overline{7}\ldots$
(e) $0.\overline{31}\ldots$ (f) $0.\overline{45}\ldots$
(g) $0.0\overline{3}\ldots$ (h) $0.1\overline{3}\ldots$

4. Let a/b and c/d denote arbitrary rational numbers. Give a convincing argument that:

(a) $\frac{a}{b} + \frac{c}{d}$ is a rational number.

(b) $\frac{a}{b} - \frac{c}{d}$ is a rational number.

(c) $\frac{a}{b} \cdot \frac{c}{d}$ is a rational number.

(d) $\frac{a}{b} \div \frac{c}{d}$ (when $c \neq 0$) is a rational number.

7-5 THE IRRATIONAL NUMBERS AND THEIR RELATION TO THE RATIONAL NUMBERS

The rational numbers form an important subset of the real numbers, but there are many real numbers that are not rational numbers. Such numbers are called **irrational** numbers. Here are a few examples of irrational numbers:

$$\sqrt{2}, \quad \sqrt[3]{10}, \quad \pi, \quad 0.101001000100001\ldots$$

How does one test to see whether a given number is rational or irrational? In view of the results stated in Section 7-4, either one of two tests can be applied:

(1) Can the number be expressed as the quotient of two integers? If this is impossible, then the number is irrational.

(2) Is the decimal expansion of the number either a terminating decimal or an infinite repeating decimal? If it is neither of these, then the number is irrational.

It is not always easy to apply these tests to a given number. Perhaps the reader knows how to prove that $\sqrt{2}$ is an irrational number. This proof is fairly easy to make. It is much harder to prove that π, which denotes the ratio of the circumference of an arbitrary circle to the diameter of the same circle, is an irrational number. In the case of the infinite decimal

$$0.101001000100001\ldots$$

one can say immediately that this number is irrational, because the decimal has a pattern that is definitely not a repeating pattern. Similarly, the infinite decimal

$$0.12112111211112\ldots$$

is an irrational number because the sequence of digits has a pattern that is not a repeating pattern.

The reader of this book does not need to know a great deal about irrational numbers, but he does need to know the following:

(1) There are many real numbers that cannot be expressed as the quotient of two integers. These numbers are called irrational numbers; the decimal expansion of each is an infinite decimal that is non-repeating.

(2) On a number line there are points that have irrational number coordinates as well as points that have rational number coordinates.

(3) Every irrational number can be approximated to any desired degree of accuracy by a rational number.

We shall now illustrate the third of these facts. The real number $\sqrt{2}$ is an irrational number: Its decimal expansion is an infinite non-repeating decimal.

$$\sqrt{2} = 1.4142\ldots$$

There is no rational number that is equal to $\sqrt{2}$, but there are rational numbers that are excellent approximations to $\sqrt{2}$. Here is a sequence of terminating decimals (and therefore a sequence of rational numbers) such that successive numbers of the sequence are better and better approximations to $\sqrt{2}$. Opposite each number there is a statement indicating how close the approximation is.

Approximation to $\sqrt{2}$	*Maximum Possible Error of the Approximation*
1	$\lvert\sqrt{2} - 1\rvert < 1$
1.4	$\lvert\sqrt{2} - 1.4\rvert < 0.1$
1.41	$\lvert\sqrt{2} - 1.41\rvert < 0.01$
1.414	$\lvert\sqrt{2} - 1.414\rvert < 0.001$
1.4142	$\lvert\sqrt{2} - 1.4142\rvert < 0.0001$
. . .	. . .

Given any small positive number ϵ (no matter how small) there is a rational number r such that

$$|\sqrt{2} - r| < \epsilon.$$

It should be emphasized that $\sqrt{2}$ is a different number from the rational number 1.414 and, in fact, $\sqrt{2}$ is different from *each* rational number. But

$$\sqrt{2} \approx 1.414$$

where the symbol $\approx$ means "is approximately equal to."

As a second example of approximating an irrational number by rational numbers, consider the irrational number

$$\pi = 3.14159. . .$$

The decimal expansion of π is an infinite, non-repeating decimal. Here is a sequence of terminating decimals such that successive members of the sequence are better and better rational number approximations to π:

Approximation to π	*Maximum Possible Error of the Approximation*
3	$\lvert\pi - 3\rvert < 1$
3.1	$\lvert\pi - 3.1\rvert < 0.1$
3.14	$\lvert\pi - 3.14\rvert < 0.01$
3.141	$\lvert\pi - 3.141\rvert < 0.001$
3.1415	$\lvert\pi - 3.1415\rvert < 0.0001$
3.14159	$\lvert\pi - 3.14159\rvert < 0.00001$
. . .	. . .

On a number line each geometric point has a unique real number as its coordinate. Some of the points have rational numbers as coordinates; others have irrational numbers as coordinates. Let P denote a point that has an irrational number as its coordinate. Let ϵ denote any small positive number such as 0.000001 or 0.00000000001. Then there is a point P^* with a *rational* number as its coordinate such that

$$d(P, P^*) < \epsilon.$$

This property is so important that it is given a special name: We say that the points with rational numbers as coordinates are **dense** on a number line. The points with rational numbers as coordinates do not "fill" the line, but they *are* numerous enough and they are so distributed that an arbitrary point on a number line either has a rational number as its coordinate itself or is not far away from a point that *does* have a rational number as its coordinate. The fact that on a number line the points with rational numbers as coordinates are dense is a consequence of the fact that between any two rational numbers there is another rational number. (See Section 1-2.)

To conclude this discussion of irrational numbers, we shall state some useful facts concerning the "nth root" of a non-negative number.

If p denotes a non-negative number, then there is a unique non-negative number, called the *square root* of p and denoted by $\sqrt{p}$, such that $(\sqrt{p})^2 = p$. Also, if p is a non-negative number and n is a positive integer, there is a unique non-negative number, called the *nth root of p* and denoted by $\sqrt[n]{p}$, such that $(\sqrt[n]{p})^n = p$. For some values of n and p the nth root of p is a rational number. For example,

$$\sqrt[3]{8} = 2$$

$$\sqrt[4]{81} = 3$$

For other values of n and p it can be shown that the nth root of p is an irrational number. The four numbers that follow are all known to be irrational numbers: $\sqrt{5}$, $\sqrt[4]{10}$, $\sqrt[3]{9}$, and $\sqrt[10]{2}$.

The following theorem (which we state without proof) is useful in identifying certain numbers as rational or irrational.

Theorem 7-5: Let n and N each denote positive integers. Then $\sqrt[n]{N}$ is a rational number only if $\sqrt[n]{N}$ is an integer.

For example, $\sqrt{17}$ is an irrational number, since there is no *integer* whose square is 17.

7-5 EXERCISES

1. Which of these numbers are irrational? Explain.

(a) 3.1415 (b) $0.\overline{23}\ldots$

(c) $\sqrt{1}$ (d) 0

(e) $\sqrt{3}$ (f) $\sqrt[4]{30}$

(g) $\sqrt[5]{1024}$ (h) $0.1\overline{34}\ldots$

2. Let z denote an arbitrary irrational number and let r denote an arbitrary non-zero rational number. Show that:

(a) $r + z$ is an irrational number.

(b) $r \cdot z$ is an irrational number.

3. Show by example that:

(a) The sum of two irrational numbers need not be an irrational number.

(b) The product of two irrational numbers need not be an irrational number.

4. Find a rational number r such that $|\sqrt{3} - r| < 0.1$.

5. Find a rational number r such that $|4\pi - r| < 0.01$.

6. Write each of these rational numbers in a simpler form:

(a) $\sqrt{49}$ (b) $\sqrt{0.49}$

(c) $\sqrt[3]{64}$ (d) $\sqrt[5]{32}$

(e) $\sqrt[5]{\frac{1}{32}}$ (f) $\sqrt{0}$

(g) $\sqrt{0.01}$ (h) $\sqrt[3]{0.001}$

7. On a number line let P denote the point that has coordinate $2\sqrt{2}$. Find the coordinate of some point P^* such that the coordinate is a rational number and $d(P,P^*) < 0.001$.

7-6 EXPONENTIAL FUNCTIONS THAT HAVE THE REAL NUMBERS AS THEIR DOMAIN

Consider the function f with domain all integers and rule $f(n) = 2^n$. The graph is an infinite set of discrete points. A few of these points are shown in Figure 7-1.

There is a natural tendency to want to connect these discrete points with some sort of a curve. Is it possible to extend f to the domain of all real numbers? The answer is yes, this can be done in

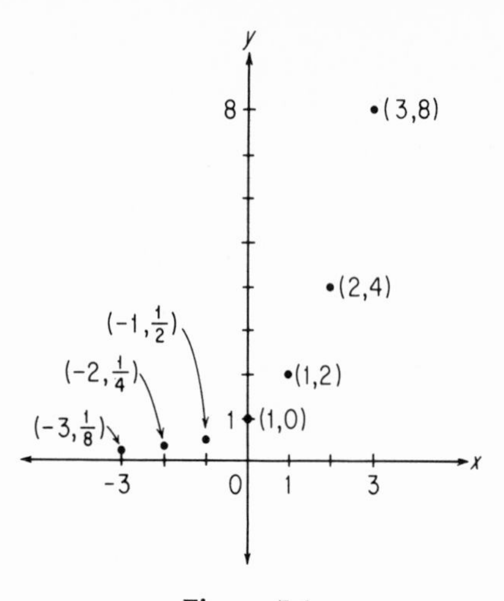

Figure 7-1

many ways. The graphs of two such extensions are shown in Figure 7-2.

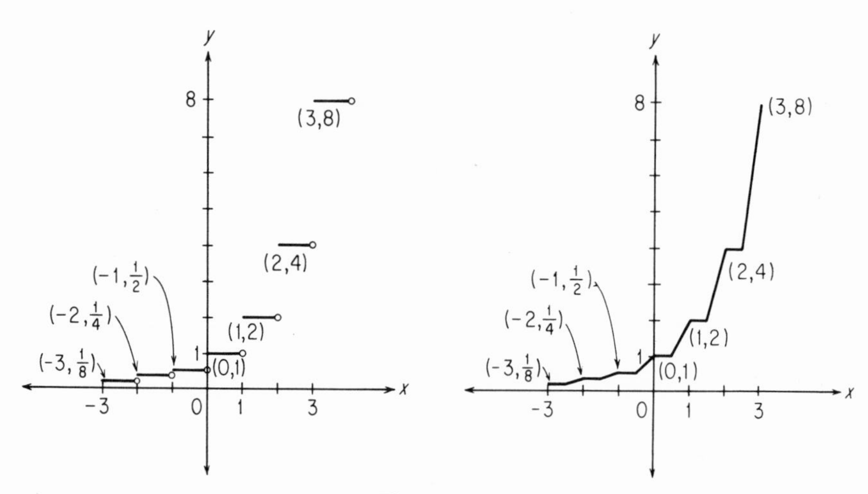

Figure 7-2

The functions whose graphs are shown in Figure 7-2 are extensions of $f(n) = 2^n$ to the domain of the real numbers, but neither of these graphs is a smooth curve without gaps and without corners. Intuition suggests that there must be some especially useful way of defining an extension of f; some way so that the graph of the extension will be a smooth curve with no gaps and no corners; some way of making the definition so that the basic rules of operation will be the same for non-integral exponents as they are for integral exponents.

There is such a definition of 2^x, for all real numbers x, whether an integer or not. This definition leads to an exponential function that has the set of *all real numbers* as its domain. The graph of this function is shown in Figure 7-3.

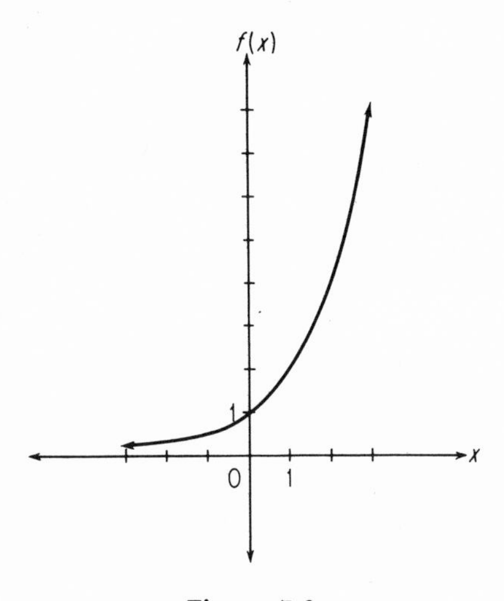

Figure 7-3

To generalize, let a denote any positive number different from 1; $f(x) = a^x$ has already been defined for each integer x. The function f can be uniquely extended to the domain of real numbers to form a new function F, with the rule also written $F(x) = a^x$, and with all the following properties:

(1) The graph of F is a smooth curve with no gaps and no corners.
(2) The function is one-to-one.
(3) The range of F is all positive numbers.
(4) If $a > 1$ the function is increasing; if $a < 1$ the function is decreasing.
(5) $F(x + y) = F(x) \cdot F(y)$.
(6) $F(x - y) = F(x)/F(y)$.
(7) $F(xy) = [F(x)]^y = [F(y)]^x$.

Note that the last three properties can be translated to read:

(5) $a^{x+y} = a^x \cdot a^y$.
(6) $a^{x-y} = a^x/a^y$.
(7) $a^{xy} = (a^x)^y = (a^y)^x$.

These are the three properties that are used so frequently in working with integral exponents. The same properties are retained when the definition of exponents is extended to include all real numbers—indeed, the definition is constructed with these properties in mind.

By now the reader must be wondering why he has not been told *how* the definition of a^x is made when x is not an integer. The reason is that the complete statement of this definition, together with the justification of the seven properties just listed, is long and tedious. Probably most readers of this book prefer to accept these facts on an intuitive basis and spend the available time getting familiar with basic properties and uses of exponential functions. The student who is interested in studying the matter more deeply is referred to [8] and [11] of the bibliography.

7-6 EXERCISES

1. Each of the function rules listed here defines an exponential function that has the set of all real numbers as its domain. Sketch the graph by plotting a few points and then drawing a smooth curve through them.

(a) $y = 3^x$ (b) $y = (\frac{1}{3})^x$

(c) $y = 10^x$ (d) $y = (0.1)^x$

(e) $y = (1.08)^x$ (See table of Section 7-3.) (f) $y = \left(\frac{1}{1.08}\right)^x$

[Hint: $y = (1.08)^{-x}$]

2. The rules given here are closely related to $y = 2^x$. Sketch the graph of each function by deciding exactly what the relationship is. In every case take the domain to be the set of all real numbers.

(a) $y = -2^x$ (b) $y = 2^{-x}$

(c) $y = -2^{-x}$ (d) $y = 2^{x-3}$

(e) $y = 2^{x+3}$ (f) $y = 3 + 2^x$

(g) $y = -4 + 2^x$ (h) $y = 2^{2x}$

(i) $y = 2^{|x|}$ (j) $y = 2^{-|x|}$

7-7 RATIONAL NUMBERS AS EXPONENTS

We have said that if $a > 0$ then a^x has meaning for *every* real number x. Also, one of the seven basic properties given in Section 7-6 is that

$a^{xy} = (a^x)^y$ for every choice of x and y. This property leads to an important conclusion about the meaning of a^x when x is a *rational* number, that is, a number that can be written in the form m/n, where m and n are integers.

Suppose that n denotes a positive integer. Then $(a^{1/n})^n = a^1 = a$. Therefore, $a^{1/n}$ is that unique positive number whose nth power is a. This means $a^{1/n} = \sqrt[n]{a}$. (See Section 7-5.) For example,

$$7^{1/3} = \sqrt[3]{7}$$
$$19^{1/5} = \sqrt[5]{19}$$
$$1000^{1/3} = \sqrt[3]{1000}.$$

Also, if m denotes any integer,

$$a^{m/n} = (a^{1/n})^m = (\sqrt[n]{a})^m.$$

Thus

$$5^{3/2} = (\sqrt{5})^3$$
$$11^{2/3} = (\sqrt[3]{11})^2$$
$$57^{-3/4} = (\sqrt[4]{57})^{-3}$$
$$8^{4/3} = (\sqrt[3]{8})^4 = 2^4 = 16.$$

The reader should not assume that if the exponent of a^x is a rational number then a^x is necessarily a rational number. Sometimes it is—but often a^x is an irrational number even though x and a are both rational numbers. For example, $5^{1/2}$ is irrational, even though 5 and $\frac{1}{2}$ are both rational numbers.

7-7 EXERCISES

1. Each of the numbers in this drill exercise has been selected so that not only is the exponent rational but the entire number is rational. Write *without* the use of exponents and in a simple form.

(a) $8^{2/3}$
(b) $8^{-2/3}$
(c) $8^{5/3}$
(d) $16^{1/4}$
(e) $16^{3/4}$
(f) $16^{-3/4}$
(g) $100^{5/2}$
(h) $100^{-5/2}$
(i) $(0.01)^{1/2}$
(j) $0.01^{5/2}$
(k) $0.01^{-5/2}$
(l) $1000^{5/3}$
(m) $256^{3/8}$
(n) $81^{1/4}$
(o) $81^{-1/4}$
(p) $(1{,}000{,}000)^{1/2}$
(q) $(10{,}000)^{1/2}$
(r) $(1{,}000{,}000)^{1/3}$

2. The solution to each of these equations (if there is a solution) is a rational number. Solve.

(a) $100^x = 1000$
(b) $100^x = 0.1$
(c) $1000^x = 10$
(d) $8^x = 128$
(e) $8^x = 0$
(f) $8^x = -8$
(g) $9^t = 27$
(h) $9^s = 243$
(i) $(\sqrt{5})^x = 125$
(j) $(\sqrt[3]{5})^x = 25$
(k) $(\sqrt[5]{7})^x = 49$
(l) $100^x = 10{,}000{,}000$
(m) $16^x = 32$
(n) $7^x = 1$
(o) $7^x = (\sqrt{7})^5$
(p) $(\sqrt{7})^x = 49$

3. Solve each equation.

(a) $2^{x-1} = 8$
(b) $2 \cdot 3^x = 54$
(c) $\dfrac{4^x}{3} = \dfrac{16}{3}$
(d) $2 \cdot 7^x + 1 = 15$
(e) $5^x + 1 = 26$
(f) $3^{2x+1} = 81$
(g) $3^x \cdot 2^x = 36$
(h) $3^x = \frac{1}{27}$
(i) $15 - 2^{-x} = 7$
(j) $2 \cdot 6^x + 3 \cdot 6^x = 30$
(k) $4^{x/2} = 64$
(l) $4^{-x} + 7 = 15$
(m) $\dfrac{27^x + 6}{5} = 3$
(n) $\left(\dfrac{1}{2}\right)^x \cdot 8^x = 16$

7-8 THE FUNCTION $F(x) = 10^x$, DOMAIN: THE SET OF REAL NUMBERS

Let F be the function that has the set of all real numbers as its domain and has the rule $F(x) = 10^x$. This function is increasing, and its range is all positive numbers. In the family of exponential functions (Section 7-6) F is especially important; our numeral system is a *decimal* system, so powers of 10 are singled out for extra attention.

In order to know F well, we need to have the answers to questions of two types:

(1) What is $F(x)$ for an arbitrary real number x?

(2) What element of the domain is mapped onto an arbitrary positive number Q?

We know that every question of the first type has a unique answer, because F is a function that has the set of all real numbers as its domain; every question of the second type has a unique answer because F is a *one-to-one* function with the set of all positive numbers as its range.

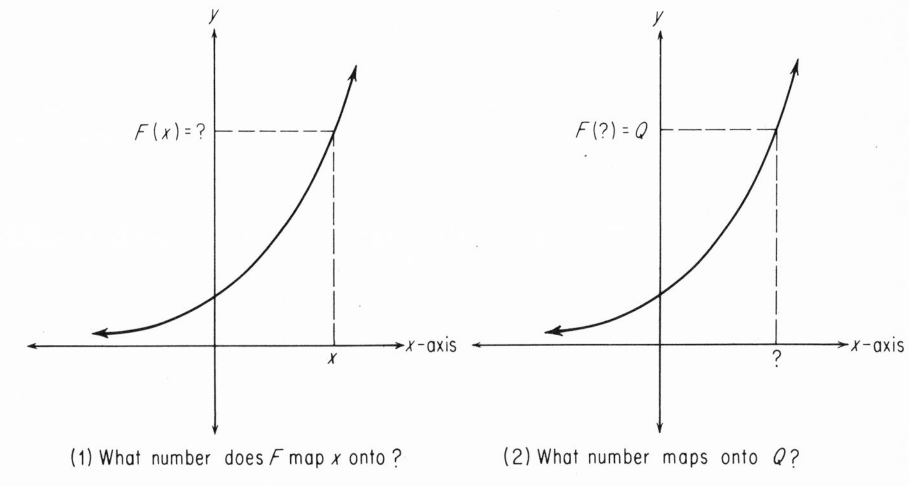

Figure 7-4

Questions of the first type have familiar answers when x is an integer:

$$F(-2) = 10^{-2} = 0.01 \qquad F(1) = 10^1 = 10$$
$$F(-1) = 10^{-1} = 0.1 \qquad F(2) = 10^2 = 100$$
$$F(0) = 10^0 = 1 \qquad F(3) = 10^3 = 1000$$

But what about $F(x)$ when x is not an integer?

$$F(-1.5) = 10^{-1.5} \qquad F(\sqrt{2}) = 10^{\sqrt{2}}$$
$$F(-0.7) = 10^{-0.7} \qquad F(2.78) = 10^{2.78}$$
$$F(0.6) = 10^{0.6} \qquad F(3.71) = 10^{3.71}$$
$$F(1.43) = 10^{1.43} \qquad F(\pi) = 10^{\pi}$$

It may seem that F is difficult to work with in one respect: Even though x is a rational number, $F(x)$ may be an *irrational* number. The reader should not feel that every irrational number *must* be approximated by a rational number in decimal form before it can be used. For every real number x, 10^x denotes a unique real number; it is a member of the system of real numbers and therefore obeys all the laws of that system. Moreover, numbers written in the form 10^x can be added, subtracted, multiplied, divided, and compared for size *without changing from the exponential form*. One of the things the reader should seek as he studies this chapter is a feeling of being "at home" with the exponential form of numbers. For example, we know

$$10^{1.5} < 10^{1.6}$$

because F is an increasing function and $1.5 < 1.6$. Also we know

$$10^{1.5} \times 10^{1.6} = 10^{3.1}$$

and

$$10^{1.5} \div 10^{1.6} = 10^{-0.1}$$

The numbers that occur in many practical problems can be left in exponential form throughout the solution of the problem—there is no need to change the form.

However, in many other applications of mathematics we do need a ready method for getting a decimal approximation to a number written in the form 10^x—and vice versa. Extensive tables have been developed to make it easy to do this. Table 1 in the Appendix is brief, but it is sufficient for our needs. More extensive tables are available in many handbooks.

Table 1 is usually called a **logarithmic table**; it could just as well be called an **exponential table.** It gives approximations for 10^x for 900 different values of x—numbers that are spaced between 0 and 1. Table 1 could be rewritten so that it looks like this:

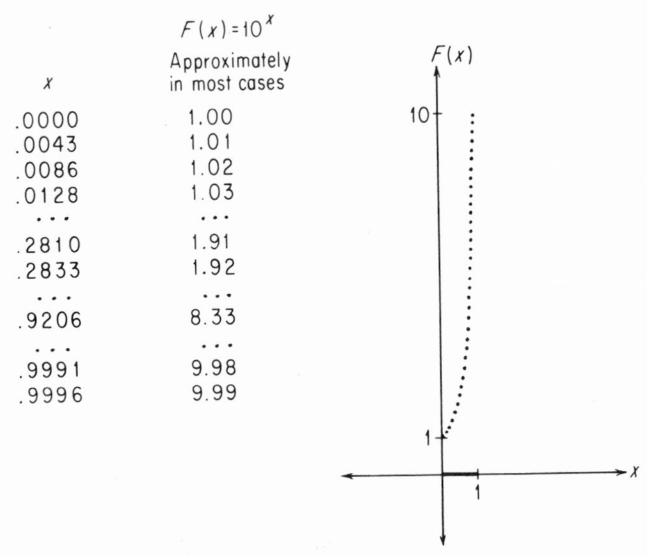

x	$F(x) = 10^x$ Approximately in most cases
.0000	1.00
.0043	1.01
.0086	1.02
.0128	1.03
...	...
.2810	1.91
.2833	1.92
...	...
.9206	8.33
...	...
.9991	9.98
.9996	9.99

Figure 7-5

This form of the table might make it easier for the beginner to use, but the full table would take up so many pages that it is impractical to print it this way.

As has been mentioned, Table 1 gives approximations to $F(x)$ for 900 different values of x—*all between 0 and 1*. This immediately suggests two questions.

Question 1: How can the table be used to approximate $F(x) = 10^x$ when x is between 0 and 1 but is *not* one of the 900 numbers of the table?

Answer: Pick the entry in the table that is closest to x. To improve on this use linear interpolation. (See Section 3-6.)

Question 2: How can the table be used to approximate $F(x) = 10^x$ when x is a number not on the interval from 0 to 1?

Answer: Write the number 10^x in the form $10^y \times 10^n$, where n is an integer and $0 < y < 1$. Then use Table 1 to get a decimal approximation for 10^y. When this is obtained you will have an approximation for 10^x that is written in scientific notation. (Observe that if y meets the condition $0 < y < 1$, then 10^y is a number between 1 and 10.) A few examples will clarify this better than anything else.

EXAMPLE 1: Use Table 1 to get a decimal approximation for $10^{0.5527}$.

The number 0.5527 is found in Table 1 in Row 3.5 under Column 7. This means that $10^{0.5527} \approx 3.57$. It should be emphasized that for every number y such that $0 < y < 1$, then $1 < 10^y < 10$.

EXAMPLE 2: Use Table 1 to get an approximation (written in scientific notation) for $10^{3.6031}$.

$$\begin{aligned} 10^{3.6031} &= 10^{0.6031} \times 10^3 \\ &\approx 4.01 \times 10^3 \end{aligned}$$

EXAMPLE 3: Use Table 1 to get a decimal approximation for $10^{-0.6003}$.

$$\begin{aligned} 10^{-0.6003} &= 10^{0.3997} \times 10^{-1} \\ &\approx 2.51 \times 10^{-1} \\ &= 0.251 \end{aligned}$$

Note that in order to use this table to approximate 10^x, where x is a *negative* number, we must first write x as a *negative integer* plus a *positive number between 0 and 1*. This can always be done.

EXAMPLE 4: Use Table 1 to get a decimal approximation for $10^{-3.1029}$.

$$\begin{aligned} 10^{-3.1029} &= 10^{0.8971} \times 10^{-4} \\ &\approx 7.89 \times 10^{-4} \\ &= 0.000789 \end{aligned}$$

EXAMPLE 5: Use Table 1 to get a decimal approximation for $10^{0.5492}$.

The number 0.5492 is between two consecutive entries of Table 1. Use linear interpolation, as explained in Section 3-6.

$$\begin{aligned} 10^{0.5492} &\approx 3.54 + \frac{3.55 - 3.54}{0.5502 - 0.5490}(0.5492 - 0.5490) \\ &\approx 3.542 \end{aligned}$$

EXAMPLE 6: Use Table 1 to get a decimal approximation for $10^{\sqrt{2}}$.

Since $\sqrt{2}$ is an irrational number, the first thing to do is to get a rational number approximation to $\sqrt{2}$; then use Table 1.

$$\begin{aligned} 10^{\sqrt{2}} &\approx 10^{1.4142} \\ &= 10^{0.4142} \times 10^{1} \\ &\approx 2.595 \times 10 \\ &= 25.95 \end{aligned}$$

Linear interpolation is used in this example also.

Not only can Table 1 be used to give approximate answers to such questions as "What is $F(1.5)$?" It can also be used to give approximate answers to the second type of question about F: For what number x is $F(x) = Q$, where Q is an arbitrary positive number?

You have known for a long time that certain numbers could be written as powers of 10: for example, $100 = 10^2$, $1000 = 10^3$, and $0.001 = 10^{-3}$. You have observed that it is especially easy to multiply and divide numbers in this form. But many numbers are not *integral* powers of 10. There is no integer n such that $17 = 10^n$, or such that $50 = 10^n$. But with the extended definition of exponents, there is a unique real number x such that $17 = 10^x$, and a unique real number y such that $50 = 10^y$. Given any positive number Q, there is a unique real number x such that $Q = 10^x$.

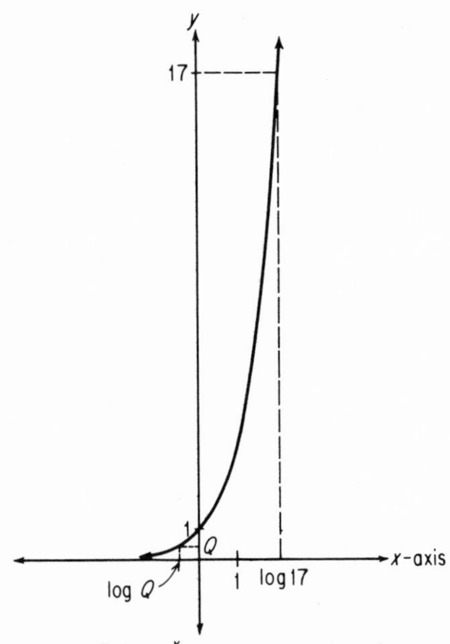

$F(x) = 10^x$, domain: the real numbers

Figure 7-6

The real number x such that $17 = 10^x$ is given a special name; it is called the "logarithm of 17 to the base 10," and this is usually abbreviated "log 17." (See Figure 7.6.)

$$17 = 10^{\log 17}$$

Similarly, the number y such that $50 = 10^y$ is called "log 50."

$$50 = 10^{\log 50}$$

In general, if Q is a positive number, then the unique real number x such that $Q = 10^x$ is called log Q. For every positive number Q,

$$Q = 10^{\log Q}$$

Table 1 may be used to find a rational number approximation to log Q for each positive number Q. In approximating log Q the first thing to do is to write Q in scientific notation.

EXAMPLE 7: Approximate 256 by a number in the form 10^x, where x is a rational number.

$$\begin{aligned} 256 &= 2.56 \times 10^2 \\ &\approx 10^{0.4082} \times 10^2 \\ &= 10^{2.4082} \end{aligned}$$

The number 0.4082 was found by looking in row 2.5 under Column 6 of Table 1.

EXAMPLE 8: Approximate 3,570,000 by a number in the form 10^x, where x is a rational number.

$$\begin{aligned} 3{,}570{,}000 &= 3.57 \times 10^6 \\ &\approx 10^{0.5527} \times 10^6 \\ &= 10^{6.5527} \end{aligned}$$

EXAMPLE 9: Approximate 0.00001675 by a number in the form 10^x, where x is a rational number.

$$\begin{aligned} 0.00001675 &= 1.675 \times 10^{-5} \\ &\approx 10^{0.2240} \times 10^{-5} \\ &= 10^{-5+0.2240} \\ &= 10^{-4.7760} \end{aligned}$$

Linear interpolation was used in expressing 1.675 as $10^{0.2240}$ approximately.

EXAMPLE 10: Use Table 1 to approximate log 1450.

We know $1450 = 10^{\log 1450}$, by definition. So this question is very much like those in Examples 7, 8, and 9.

$$\begin{aligned} 1450 &= 1.450 \times 10^3 \\ &= 10^{0.1614} \times 10^3 \\ &= 10^{3.1614} \end{aligned}$$

Therefore,

$$\log 1450 \approx 3.1614.$$

EXAMPLE 11: Use Table 1 to approximate log 0.00167. We know $0.00167 = 10^{\log 0.00167}$.

$$\begin{aligned} 0.00167 &= 1.67 \times 10^{-3} \\ &\approx 10^{0.2227} \times 10^{-3} \\ &= 10^{-3+0.2227} \\ &= 10^{-2.7773} \end{aligned}$$

Therefore,

$$\log 0.00167 \approx -2.7773.$$

7-8 EXERCISES

1. Write each of these numbers in the form $n + x$, where n is an integer and $0 < x < 1$.

(a) 4.586 (b) -4.586
(c) -3.781 (d) -0.0061
(e) -10.98 (f) -1.012

2. Let $F(x) = 10^x$, domain: the real numbers. Use Table 1 to get decimal approximations to three significant digits for:

(a) $F(0.6042)$ (b) $F(3.6875)$
(c) $F(0.9238)$ (d) $F(5.9238)$
(e) $F(0.8169)$ (f) $F(1.8169)$
(g) $F(1.6)$ (h) $F(2.8100)$
(i) $F(-4.2)$ (j) $F(-0.04)$
(k) $F(-1.78)$ (l) $F(-5.0240)$

3. Give the appropriate logarithmic name to the solution of each of these equations. Then give the solution without the use of the word logarithm. Table 1 is not needed.

(a) $10^x = 100$ (b) $10^x = 1000$
(c) $10^x = 1{,}000{,}000$ (d) $10^x = 1$
(e) $10^x = 1{,}000{,}000{,}000$ (f) $10^x = 0.1$
(g) $10^x = 0.01$ (h) $10^x = 0.001$
(i) $10^x = 10^{50}$ (j) $10^x = 10^{-40}$

4. Give the appropriate logarithmic name to the solution of each of these equations. Then use Table 1 to give a decimal approximation to each solution.

(a) $10^x = 50$ (b) $10^x = 200$
(c) $10^x = 1500$ (d) $10^x = 2400$
(e) $10^x = 1.8$ (f) $10^x = 0.18$
(g) $10^x = 0.018$ (h) $10^x = 0.00018$
(i) $10^x = 452$ (j) $10^x = 5{,}680{,}000$
(k) $10^x = 0.000897$ (l) $10^x = 99$

7-9 USING EXPONENTS TO SIMPLIFY CALCULATIONS

Many tedious calculations can be made relatively easy by first expressing some of the numbers involved in the form 10^x and then using the basic properties:

$$10^x 10^y = 10^{x+y}$$

$$\frac{10^x}{10^y} = 10^{x-y}$$

$$(10^x)^y = 10^{xy}$$

EXAMPLE 1: Use Table 1 to approximate $\sqrt[3]{3450}$ by a number in the form 10^x.

$$\begin{aligned}\sqrt[3]{3450} &= (3450)^{1/3} \\ &\approx (10^{3.5378})^{1/3} \\ &= 10^{1.1793}\end{aligned}$$

This is the answer in exponential form, as required. If the answer is wanted in decimal form also, use Table 1 again:

$$\begin{aligned}10^{1.1793} &= 10^{0.1793} \times 10^1 \\ &\approx 1.511 \times 10^1 \\ &= 15.11\end{aligned}$$

Therefore,

$$\sqrt[3]{3450} \approx 15.11.$$

Linear interpolation was used in this example.

EXAMPLE 2: Use Table 1 to get a decimal approximation (with three significant digits) for $257/\sqrt[3]{3450}$. Notice that $\sqrt[3]{3450}$ is the number that was approximated in exponential form in Example 1.

$$\begin{aligned}\frac{257}{\sqrt[3]{3450}} &\approx \frac{10^{2.4099}}{10^{1.1793}} \\ &= 10^{1.2306} \\ &= 10^{0.2306} \times 10^1 \\ &\approx 1.70 \times 10^1 \\ &= 17.0\end{aligned}$$

EXAMPLE 3: Use Table 1 to get an approximation for $3^{10} \times 2^{0.1}$. Express the answer in scientific notation with three significant digits.

$$\begin{aligned} 3^{10} \times 2^{0.1} &\approx (10^{0.4771})^{10} \times (10^{0.3010})^{0.1} \\ &= 10^{4.771} \times 10^{0.0301} \\ &= 10^{4.8011} \\ &= 10^{0.8011} \times 10^4 \\ &\approx 6.33 \times 10^4 \end{aligned}$$

7-9 EXERCISES

Use Table 1 to approximate these numbers. Express the final answer in decimal form or in scientific notation with three significant digits.

1. $\sqrt{472}$
2. $\sqrt{0.023}$
3. $\sqrt[3]{1290}$
4. 2^{20}
5. 7^{10}
6. $(0.019 \times 4.82)/3^{10}$
7. $\sqrt[3]{18} \times \sqrt{95}$
8. $450/\sqrt{68{,}300}$
9. $4^5/5^{10}$
10. $(0.3)^{10}$

7-10 EXPONENTIAL GROWTH AND DECAY

Suppose a positive quantity Q is increasing (or decreasing) with time and the instantaneous rate of its increase (or decrease) is directly proportional to Q. Then it can be shown that the rule for $Q(t)$ has the form $Q(t) = k \cdot a^t$, where k and a are positive constants.

Many quantities that scientists are interested in behave in approximately this fashion; $Q(t) = k \cdot a^t$ is the mathematical model that the scientist uses to predict Q at a given time. Two prominent examples are (1) the growth of certain populations, and (2) radioactive decay.

We shall call any function whose rule has the form $f(x) = k \cdot a^x$ (where $a > 0$ and $a \neq 1$) an *exponential* function. This means that the family of exponential functions is a two-parameter family—the parameters being a and k; any specific choice of a and k yields a specific member of the family.

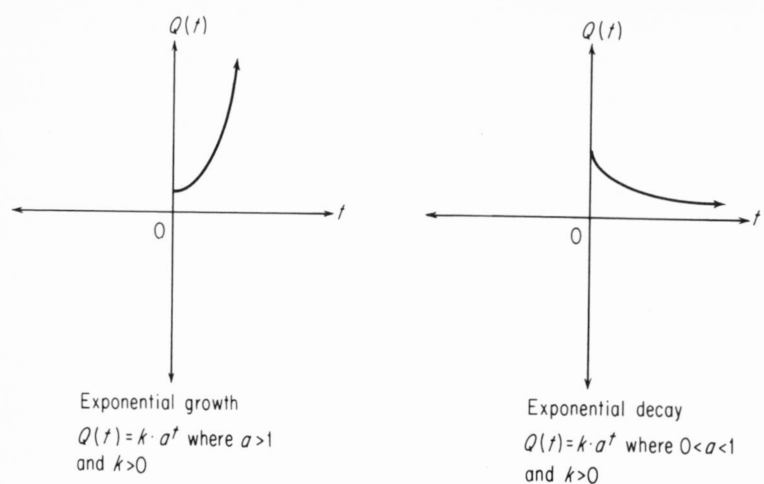

Figure 7-7

EXAMPLE: A scientist has reason to believe that a certain quantity Q is growing exponentially: $Q(t) = k \cdot a^t$, where k and a denote positive numbers that are not yet determined and t denotes time measured in some appropriate unit. The scientist finds that $Q(3) = 40$ and $Q(5) = 160$. Determine the parameters k and a. How large will Q be when $t = 10$?

Solution: Direct substitution yields

$$k \cdot a^3 = 40$$

and

$$k \cdot a^5 = 160.$$

Both these conditions must be satisfied by k and a. Substitute 40 in place of $k \cdot a^3$ in the second equation:

$$k \cdot a^5 = 160$$

$$\Leftrightarrow (k \cdot a^3)a^2 = 160$$

$$\Leftrightarrow 40 \cdot a^2 = 160$$

$$\Leftrightarrow a^2 = 4$$

$$\Leftrightarrow a = 2 \qquad (a \text{ must be positive})$$

Substitute $a = 2$ in the equation

$$k \cdot a^3 = 40$$

to obtain $k = 5$.

The function rule is

$$Q(t) = 5 \cdot 2^t.$$

In particular,

$$Q(10) = 5 \cdot 2^{10} = 5120.$$

The next set of exercises is presented to give the reader some idea of applications of exponential functions to questions about growth or decay.

7-10 EXERCISES

1. Each of the function rules listed here defines a member of the family of exponential functions. In each case the domain is given to be the set of real numbers. Sketch the graph of each function.

(a) $y = 2 \cdot 3^x$ (b) $y = -2 \cdot 3^x$

(c) $y = 3 \cdot (\frac{1}{2})^x$ (d) $y = \frac{1}{4} \cdot 3^x$

2. (a) Suppose \$100 could be invested at 100% compounded annually. Let $A(t)$ denote the amount of this investment at any time t. Assume also that A does not increase during the year but only at the end of each year. Use the "bracket" function of Section 2-5 to write the rule $A(t)$. Sketch the graph of this function for $0 \leqq t \leqq 4$. (Note: This is not an exponential function—just a relative of the family.)

(b) Use the same set of axes as for part (a) and sketch the graph of the exponential function Q, defined by $Q(t) = 100 \cdot 2^t$ for $0 \leqq t \leqq 4$.

3. Suppose g meets all these conditions:

(1) The domain is the set of real numbers.

(2) $g(x) = k \cdot a^x$, where k and a are positive numbers yet to be determined.

(3) $g(-1) = 12$ and $g(2) = \frac{4}{9}$.

(a) Determine k and a and write the rule for g.

(b) Compute $g(4)$.

4. A scientist knows that a certain quantity Q is increasing with time (measured in some appropriate unit). He knows that $Q(0) = 2$ and $Q(1) = 4$. What should he predict for $Q(4)$ if he has reason to believe that Q is increasing according to:

(a) A linear function rule?

(b) A rule of the form $Q(t) = at^2 + c$?

(c) A rule of the form $Q(t) = k \cdot a^t$?

5. The number of bacteria in a culture at time t is given by $N(t) = N_0 \cdot 4^t$, where N_0 is some fixed positive number and t is measured in hours.

(a) What was the number present when $t = 0$?

(b) Approximately how long will it take for the number of bacteria to double?

6. Suppose the population of a certain country was two hundred million in 1940. Since that time its population has followed an exponential growth pattern, $P(t) = 2 \cdot 10^8 \cdot (1.1)^t$, where t is measured in decades. Assume that this growth pattern continues. Use Table 1 to estimate:

(a) The 1975 population.
(b) The 1980 population.
(c) The 1995 population.
(d) The 2040 population.

7. A certain radioactive substance decomposes in such a way that four-fifths of a given amount of the substance remains after one year. Assume that the mathematical model of exponential decay is appropriate: $A(t) = A_0 \cdot a^t$, where t is measured in years.

(a) What part of a given amount of the substance will remain after 4 years?
(b) Approximate the half-life of the substance.

7-11 THE FUNCTION $G(x) = \log x$, DOMAIN: THE POSITIVE NUMBERS

In Section 7-9 we discussed the function defined by $F(x) = 10^x$, domain: the real numbers. The range of F is the positive numbers.

Since F is a one-to-one function, we know it has an inverse. We shall denote this inverse function by G; its domain is the positive numbers (this is the range of F); its range is the real numbers (this is the domain of F); its graph is congruent to that of F, and is, in fact, the reflection of the graph of F in the line $\{(x, y) : y = x\}$. (See Figure 7-8.)

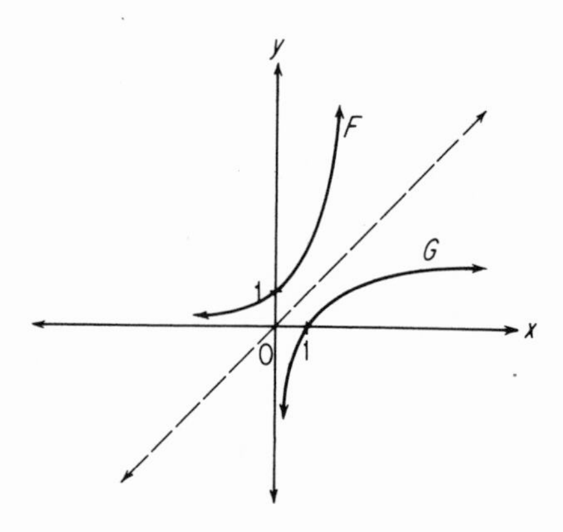

Figure 7-8

We already have a special name provided for the inverse function: $G(x) = \log x$. This is appropriate because $F(\log x) = 10^{\log x} = x$, where x denotes any positive number. The inverse function G maps x onto $\log x$. (See Figure 7-9.)

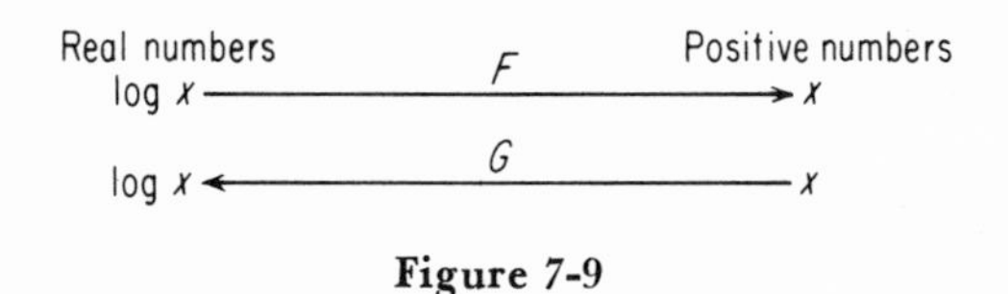

Figure 7-9

Remember that for a given positive number x, the number $\log x$ is the unique number with the property that:

$$10^{\log x} = x$$

This is often expressed as follows:

$$10^y = x \Leftrightarrow y = \log x$$

Table 1, an exponential table, is also a logarithmic table. We can use it (we have already done so) to approximate $G(x)$ for an arbitrary positive x. (See Section 7-8.)

7-11 EXERCISES

1. Let $G(x) = \log x$, domain: the positive numbers. Evaluate:

(a) $G(10)$ (b) $G(100)$
(c) $G(1000)$ (d) $G(1{,}000{,}000)$
(e) $G(1)$ (f) $G(0.01)$
(g) $G(0.001)$ (h) $G(0.1)$
(i) $G(10^{40})$ (j) $G(10^n)$

2. Use Table 1 to get decimal approximations for the following numbers. (See Section 7-8, Example 10.)

(a) $G(2)$ (b) $G(5)$
(c) $G(3.78)$ (d) $G(0.002)$
(e) $G(0.125)$ (f) $G(75)$
(g) $\log 1430$ (h) $\log 58{,}600$
(i) $\log 0.012$ (j) $\log 0.0078$

3. The number 75 is between 10 and 100. Therefore $\log 75$ is between 1 and 2. Use a similar argument and locate each of these numbers between consecutive integers.

(a) $\log 256$ (b) $\log 1897$

(c) $\log 687$ (d) $\log 89{,}765$
(e) $\log 0.89$ (f) $\log 0.0089$
(g) $\log 0.7$ (h) $\log 0.00008$

4. Sketch the graph of each function in this list after first deciding how the function is related to $G(x) = \log x$.

(a) $y = -\log x$, domain: the positive numbers.
(b) $y = \log(-x)$, domain: the negative numbers.
(c) $y = \log |x|$, domain: all numbers except 0.
(d) $y = |\log x|$, domain: all positive numbers.
(e) $y = 2 + \log x$, domain: all positive numbers.
(f) $y = \log(x - 2)$, domain: $\{x: x > 2\}$.
(g) $y = \log(x + 3)$, domain: $\{x: x > -3\}$.

7-12 A FAMILY OF LOGARITHMIC FUNCTIONS

Let $a > 0$ and $a \neq 1$. The rule $f(x) = a^x$ defines a family of exponential functions having as its domain the set of all real numbers. Each member of this family is a one-to-one function, with the range being the set of all positive numbers.

Therefore, each member of the family of exponential functions has an inverse function: the domain is the set of positive numbers. Each is known as a *logarithmic* function. The inverse of $F(x) = 10^x$ [that is, $G(x) = \log x$] is just one member of a large family. The a of the exponential function $f(x) = a^x$ is called the *base*; to distinguish among the many logarithmic functions the a is written as a subscript. The inverse of $f(x) = a^x$ is $g(x) = \log_a x$. Only in the case when $a = 10$ is the base omitted from the notation.

Exponential Functions	*Corresponding Inverses*
$y = 10^x$	$y = \log x$
$y = 2^x$	$y = \log_2 x$
$y = 3^x$	$y = \log_3 x$
$y = 7^x$	$y = \log_7 x$
$y = 13^x$	$y = \log_{13} x$
$y = (\frac{1}{2})^x$	$y = \log_{1/2} x$

The graph of each inverse function is obtained by reflecting the graph of the corresponding exponential function in the line $\{(x, y): y = x\}$. Figure 7-10 shows the graph of $f(x) = 2^x$ and its inverse $g(x) = \log_2 x$.

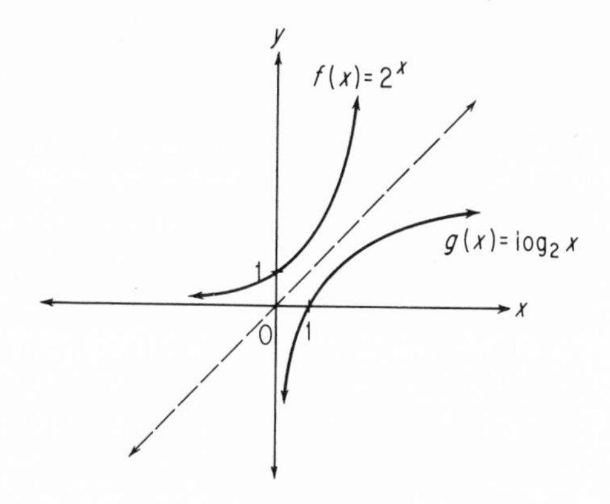

Figure 7-10

If $a > 0$ and $a \neq 1$ and if Q is some positive number, then there is a unique number x such that $Q = a^x$. This number is called "the logarithm of Q to the base a" and is abbreviated $\log_a Q$. For example, there is a number x such that $17 = 2^x$. This number x is an irrational number; its short name is "$\log_2 17$."

$$17 = 2^{\log_2 17}$$

Similarly, the number y such that $89 = 7^y$ is called the "$\log_7 89$."

$$89 = 7^{\log_7 89}$$

In general,

$$Q = a^{\log_a Q}.$$

7-12 EXERCISES

1. Sketch the graph of each of these functions and its inverse function.
(a) $f(x) = 2^x$.
(b) $g(x) = 7^x$.
(c) $h(x) = (\frac{1}{2})^x$.

2. Give an appropriate logarithmic name to the unique solution of each of these equations.

(a) $2^x = 20$
(b) $2^x = 75$
(c) $3^x = 40$
(d) $3^x = 0.001$
(e) $12^x = 15$
(f) $5^x = 75$
(g) $2^x = 16$
(h) $2^x = 100$
(i) $10^x = 100$
(j) $10^x = 1$
(k) $2^x = 1$
(l) $10^x = 1000$

3. Because 7 is between 2^2 and 2^3, $\log_2 7$ is between 2 and 3. In a similar way, locate each of these irrational numbers between two consecutive integers.

(a) $\log_2 15$
(b) $\log_2 75$
(c) $\log_2 127$
(d) $\log_2 400$
(e) $\log_2 1000$
(f) $\log_2 0.6$
(g) $\log_2 0.06$
(h) $\log_3 19$
(i) $\log_3 (\frac{1}{2})$
(j) $\log_3 45$
(k) $\log_5 79$
(l) $\log_{1/2} 43$

4. Each of these numbers is a rational number. Write each in decimal form.

(a) $\log_3 9$
(b) $\log_3 \frac{1}{3}$
(c) $\log_2 64$
(d) $\log_{1/2} 64$
(e) $\log_2 128$
(f) $\log_2 (\frac{1}{64})$
(g) $\log_4 32$
(h) $\log_8 64$
(i) $\log_{16} 64$
(j) $\log_{27} 81$
(k) $\log_{1000} 1000$
(l) $\log_{1000} 100$

8

Periodic Functions

8-1 WHAT IS A PERIODIC FUNCTION?

Before introducing a familiar example of a periodic function we state again an important theorem known as "the division algorithm." It concerns the division of an integer by an arbitrary positive integer.

Theorem: Let n denote any integer and let d denote a positive integer. Then there exist unique integers q and r such that $n = qd + r$, where $0 \leqq r < d$.

In this theorem d is called the "divisor" and r the "remainder." For any given positive integer d the integers serve as the domain of a remainder function f with function values $f(n) = r$, where $0 \leqq r < d$. The following table gives functional values of f when $d = 5$.

n	$\cdots$ -5	-4	-3	-2	-1	0	1	2	3	4	5	6 $\cdots$
$f(n)$	$\cdots$ 0	1	2	3	4	0	1	2	3	4	0	1 $\cdots$

This table reveals a periodic or cyclic property of f. The same periodic property is also suggested by the graph of f (Figure 8-1). Observe that f meets these two conditions:

(1) If a number n is in the domain, then so is $n + 5$.
(2) $f(n + 5) = f(n)$ for every integer n.

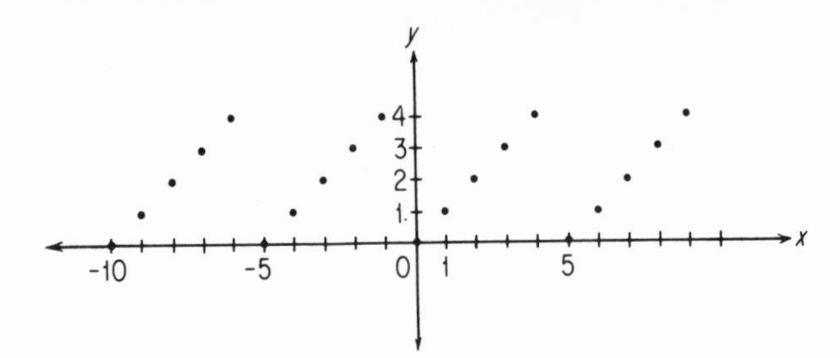

Figure 8-1

Because of property (2) the number 5 is called a **period** of f. Any positive integral multiple of 5 is also a period of f. For every integer n,

$$f(n + 10) = f(n)$$
$$f(n + 15) = f(n)$$
$$f(n + 20) = f(n)$$
$$\cdots$$

Among all the positive numbers p such that for every integer n

$$f(n + p) = f(n)$$

the number 5 is especially important because it is the smallest such number. The **fundamental period** of f is 5.

The remainder function just discussed suggests general conditions required of a function that is to be called "periodic."

Definition: Let f be a function that satisfies all of the following:
1. The domain D is a nonempty subset of the real numbers.
2. There is a positive number p such that if x belongs to D then $x + p$ belongs to D.
3. $f(x + p) = f(x)$ for each x in D.
Then f is said to be a **periodic function** with **period *p*.**

Note that if p is a period of a function, then so are $2p$, $3p$, . . . , np, ... If there is a least positive number p such that p is a period of f, then this number is known as the **fundamental period.**

The domain of a periodic function may be all real numbers; indeed, this is the domain for many of the functions discussed in this chapter. However, for other periodic functions the domain may be a proper subset of the real numbers. In any event, the domain must have the property that if p is a period of the function and if x belongs to the domain, then $x + p$ also belongs to the domain.

8-1 EXERCISES

1. Let f denote a remainder function that has the integers as its domain and has the rule $f(n) = r$, where r is the remainder obtained when n is divided by $d = 2$. This is a periodic function with fundamental period 2. What is the range of f?

(a) $f(2) = ?$ (b) $f(17) = ?$
(c) $f(101) = ?$ (d) $f(-5) = ?$
(e) $f(2n) = ?$ (f) $f(2n + 1) = ?$

2. Let the domain of g be the integers and let the rule be $g(n) = r$, where r denotes the remainder obtained when n is divided by 7.

(a) $g(8) = ?$
(b) $g(-10) = ?$
(c) $g(100) = ?$
(d) What is the range of g?
(e) What is the fundamental period of g?

3. If today is Wednesday, what day of the week will it be:

(a) 10 days from now?
(b) 100 days from now?
(c) $7n + 3$ days from now, where n is any positive integer?

4. In 1972 Christmas day falls on Monday. What day of the week will Christmas fall on in:

(a) 1973? (b) 1974?

5. Suppose it is now 10 a.m.

(a) What time of day (or night) will it be 100 hours from now?
(b) What time will it be 1000 hours from now?

6. The graphs of some functions are given below. Select those that seem to be periodic. If the function has a fundamental period, state what this is.

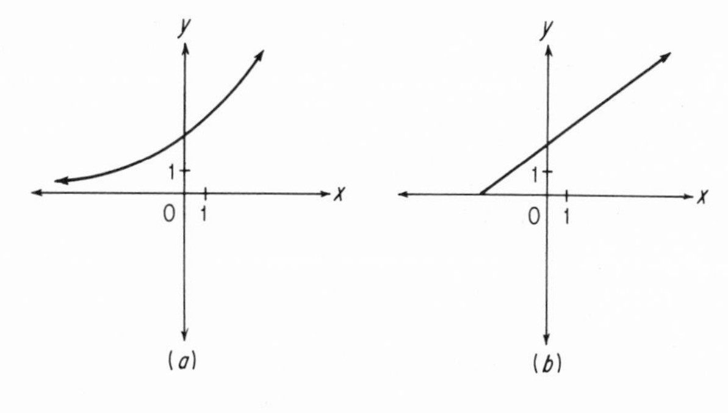

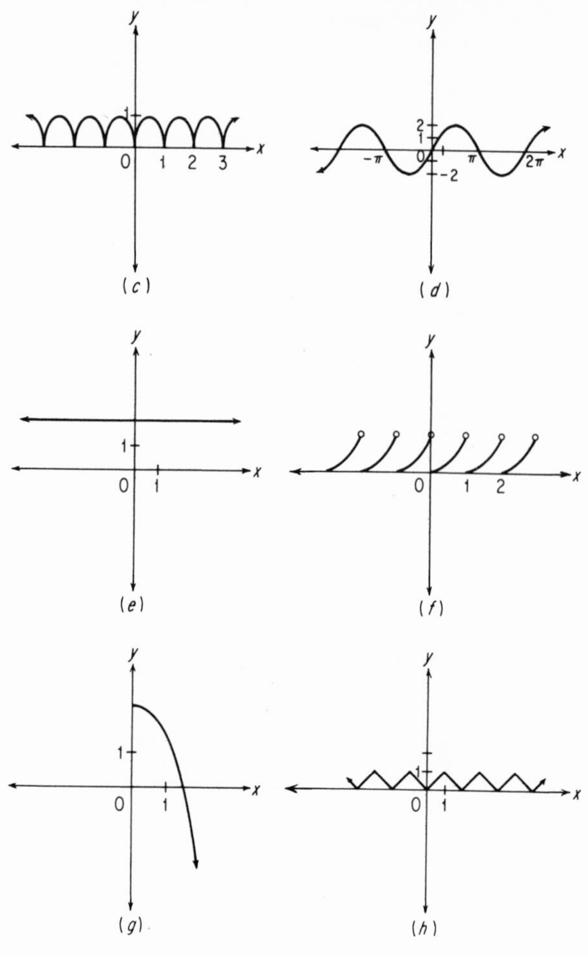

7. A function F is defined by giving the following information about it:

(1) The domain of F is the set of real numbers.
(2) F is periodic with period 2.
(3) $F(x) = x^2$ on the interval $\{x: 0 \leqq x < 2\}$.

(a) Sketch the graph of F.
(b) $F(3) = ?$
(c) $F(16) = ?$
(d) $F(-7) = ?$
(e) $F(-100.5) = ?$

8. A function G is defined by the following conditions:

(1) The domain of G is the set of real numbers.
(2) G is periodic with period 5.
(3) $G(x) = 5 - x$ on the interval $\{x: 0 \leqq x < 5\}$.

(a) Sketch the graph of G.
(b) $G(12) = ?$
(c) $G(-11) = ?$

9. A function H is defined by the following conditions:
(1) The domain of H is the real numbers.
(2) H is periodic with period 3.
(3) $H(x) = \sqrt{9 - x^2}$ on the interval $\{x: 0 \leqq x < 3\}$.
(a) Sketch the graph of H.
(b) $H(15) = ?$
(c) $H(-7) = ?$

10. A function I is defined by the following conditions:
(1) The domain of I is the real numbers.
(2) I is periodic with period 2.
(3) $I(x) = |x - 1|$ on the interval $\{x: 0 \leqq x < 2\}$.
(a) Sketch the graph of I.
(b) $I(-3) = ?$
(c) $I(19) = ?$

11. A function J has the set of all real numbers as its domain. The graph of J is the union of parabolic arcs shown here. State conditions (similar to those in Exercise 7) that would define a function whose graph would have about the same appearance as the graph of J.

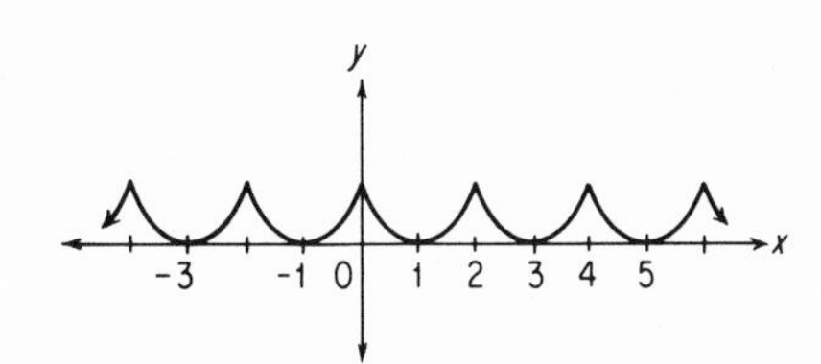

12. A function L has the set of all real numbers as its domain. The graph of L is the union of parabolic arcs shown here. State conditions (similar to those in Exercise 7) that would define a function whose graph would have about the same appearance as the graph of L.

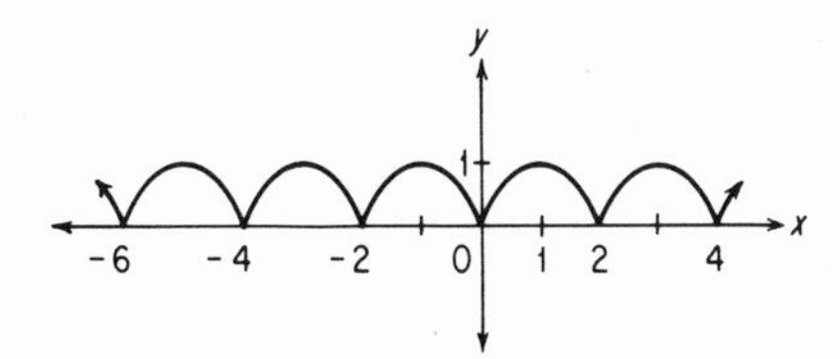

13. In Section 2-5, Exercise 4 the symbol $[x]$ is used to denote the greatest integer less than or equal to x (where x can be any real number).

(a) Does $[x + 1] = 1 + [x]$ for every x? Explain.

(b) Let $f(x) = x - [x]$, domain: the real numbers. Show that this function is periodic and that its fundamental period is 1. [Hint: Use the result of part (a).]

(c) Sketch the graph of f.

14. It is possible for a function to be a periodic function and yet not to have a *fundamental* period. Give a simple example of such a function.

15. Can a one-to-one function from the real numbers into the real numbers be a periodic function? Explain.

16. Can a function whose domain is a finite set be periodic? Explain.

8-2 SOME SQUARE FUNCTIONS

This section concerns three periodic functions that are called "square" functions because they are defined by using a certain geometric square in a coordinate plane.

In a rectangular coordinate plane consider the square with vertices at the points whose coordinates are $(1, 1)$, $(-1, 1)$, $(-1, -1)$, and $(1, -1)$. The perimeter of this square is 8 (Figure 8-2).

Let L denote the line that is parallel to the y-axis and contains the points with coordinates $(1, 1)$, $(1, 0)$, and $(1, -1)$. To each real number y there corresponds a unique point on L—the point with coordinates $(1, y)$. Also each point on L has coordinates $(1, y)$ for some number y; therefore, to each point $(1, y)$ on L there corre-

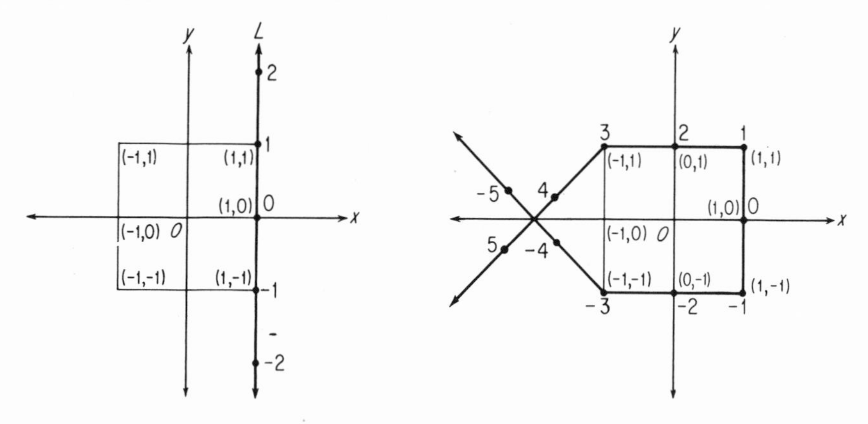

Figure 8-2

sponds the unique real number y. Thus L can be thought of as a number line, with each of its points denoted by a *single* coordinate.

Now in our imagination let us wind the number line L (both halves of it) tightly about the square, keeping the point of L with coordinate 0 at the point with coordinates $(1, 0)$ in the plane and keeping the point on L with coordinate 1 at the point with coordinates $(1, 1)$ in the plane. The positive half of the line is wound in a counter-clockwise direction; the negative half is wound in a clockwise direction. This winding is done without any stretching or contracting of L. Each point of L is mapped onto a point on the square. Therefore, each real number (the coordinate of a point on L) is mapped onto an ordered pair of real numbers (the coordinates of a point on the square). This mapping defines a function that we shall call P. The domain of this function is the set of real numbers, and the range is a set of ordered *pairs* of real numbers.

Since the perimeter of the square is 8, P is a periodic function with period 8. The function P maps each real number t onto an ordered pair (x, y) of real numbers. The function value $P(t) = (x, y)$ if and only if the point with coordinate t on the real line is mapped onto the point with coordinates (x, y) on the square when the line is wrapped around the square.

Now let C denote the function that maps t onto the x-coordinate of the ordered pair $P(t) = (x, y)$, and let S denote the function that maps t onto the y-coordinate. This defines two additional periodic functions, but whereas the range of P is a set of ordered pairs of numbers, the range of C and the range of S are sets of numbers. The following table gives functional values of P, C, and S.

t	$P(t)$	$C(t)$	$S(t)$
-4	$(-1, 0)$	-1	0
-3	$(-1, -1)$	-1	-1
-2	$(0, -1)$	0	-1
-1	$(1, -1)$	1	-1
0	$(1, 0)$	1	0
1	$(1, 1)$	1	1
2	$(0, 1)$	0	1
π	$(-1, 4 - \pi)$	-1	$4 - \pi$
3.5	$(-1, 0.5)$	-1	0.5
4	$(-1, 0)$	-1	0
4.1	$(-1, -0.1)$	-1	-0.1
5	$(-1, -1)$	-1	-1
5.7	$(-0.3, -1)$	-0.3	-1

We cannot sketch the graph of P in a coordinate plane, but we can sketch the graphs of C and S. The graph of C is shown in Figure 8-3.

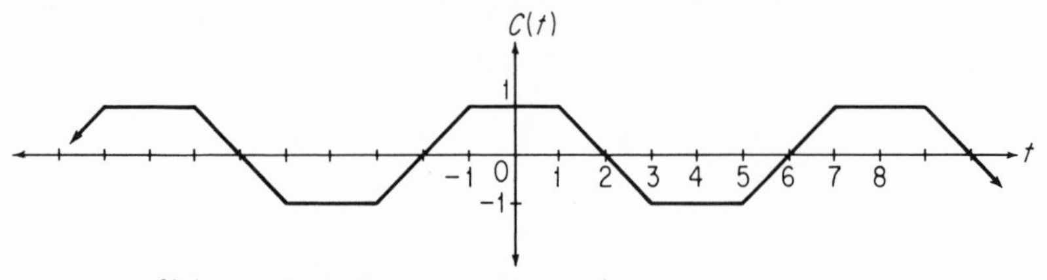

C(t) is the first coordinate of the point (x, y) when P(t)=(x, y)

Figure 8-3

The function C is also completely described as follows:

(1) The domain of C is the set of real numbers.
(2) C is a periodic function with period 8.
(3)

$$C(t) = \begin{cases} 1, & \text{if } 0 \leqq t \leqq 1 \\ 2 - t, & \text{if } 1 < t \leqq 3 \\ -1, & \text{if } 3 < t \leqq 5 \\ t - 6, & \text{if } 5 < t \leqq 7 \\ 1, & \text{if } 7 < t \leqq 8 \end{cases}$$

8-2 EXERCISES

In all these exercises the symbols P, C, and S denote the square functions defined in this section.

1. Complete the following table of functional values:

t	$P(t)$	$C(t)$	$S(t)$
10	(0, 1)		
15		1	−1
17	(1,)		1
18			
20			
100	(−1, 0)	−1	
1000	(1, 0)		
−80	(, 0)	1	
−14		0	1
−1000			
10^6			

2. What is the range of C? Of S?

3. (a) Sketch the graph of S.
(b) Do you think the graph of C is congruent to the graph of S? Explain.

4. Let f be defined by $f(x) = x + 2$ for all x. Sketch the graph of the composite function Cf.

5. Let g be defined by $g(x) = x - 4$. Sketch the graph of the composite function Cg.

6. Let h be defined by $h(t) = 2t$.
(a) Sketch the graph of the composite function Ch.
(b) What is the fundamental period of Ch?

7. Prove or disprove:
(a) C maps each rational number onto a rational number.
(b) C maps each irrational number onto an irrational number.

8-3 THE SINE AND COSINE FUNCTIONS

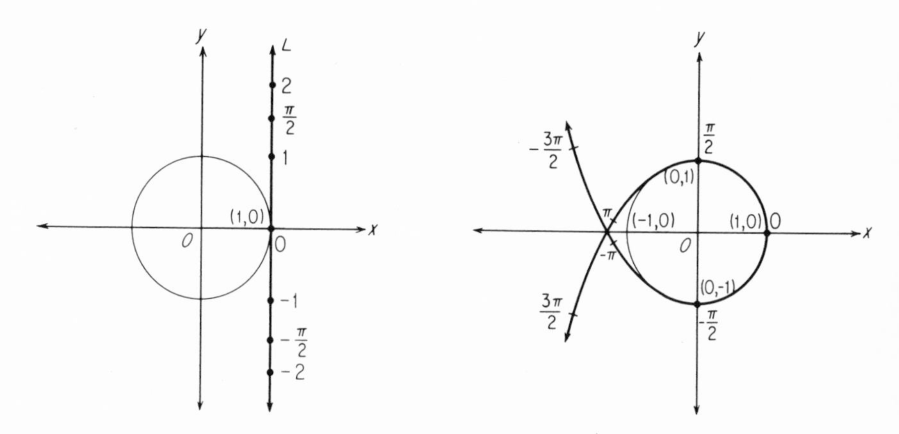

Figure 8-4

If the geometric square of Section 8-2 is replaced by a unit circle with center at the origin, three circular functions can be defined that are analogous to the three square functions. (The square functions serve primarily to introduce the circular functions—it is the latter that are really important to the reader.) The circumference of this circle is 2π. As in Section 8-2, we let L denote a real number line that is parallel to the y-axis with the point on L whose coordinate is 0 at the point on the unit circle with rectangular coordinates $(1, 0)$. Imagine winding L (both halves of it) tightly about the unit circle (Figure 8-4). The half-line that contains the point with coordinate 1 is wound in the counterclockwise direction; the half-line that contains the point with coordinate -1 is wound in the clockwise direction. The point with coordinate 0 on L is held fixed at the point with coordinates $(1, 0)$ on the plane. In the imagination L can be pictured as an infinitely long tape measure—one that is flexible (but not stretchable) and that has no thickness.

Each point on the tightly wound line is mapped onto a point on the circle, and thus each real number is mapped onto an ordered pair of real numbers. This mapping defines a function that we shall call Q. The domain of Q is the set of real numbers, and the range is a set of ordered *pairs* of real numbers. The function $Q(t) = (x, y)$ if and only if the point with coordinate t on the number line is mapped onto the point (x, y) on the circle when the line is wrapped around the circle. Since the circumference of the circle is 2π, Q is a periodic function with period 2π.

We now define two more functions called the **cosine function** and the **sine function.** The abbreviations **cos *t*** and **sin *t*** are commonly used to denote the functional values. Each of these functions has the set of real numbers as its domain; the rules for $\cos t$ and $\sin t$ are defined as follows:

$$\text{If } Q(t) = (x, y), \quad \text{then} \quad \cos t = x$$

$$\text{If } Q(t) = (x, y), \quad \text{then} \quad \sin t = y$$

How does one determine $\cos t$ for a given number t? To answer this question, first determine $Q(t)$ as defined in the preceding paragraph. This will be an ordered pair (x, y) that designates a point on the unit circle. Select the first member of the pair. This will be $\cos t$. The second member of the pair will be $\sin t$.

The cosine and sine functions have many applications in physics, particularly to the study of radio waves, sound waves, heat, and mechanics. The cosine and sine functions are basic to everything that follows in this chapter.

Like Q, both the cosine and sine functions are periodic with period 2π. The following table gives $Q(t)$, $\cos t$, and $\sin t$ for a few specially chosen values of t.

t	$Q(t)$	$\cos t$	$\sin t$
0	$(1, 0)$	1	0
$\frac{\pi}{2}$	$(0, 1)$	0	1
$-\frac{\pi}{2}$	$(0, -1)$	0	-1
π	$(-1, 0)$	-1	0
$-\pi$	$(-1, 0)$	-1	0
$\frac{3\pi}{2}$	$(0, -1)$	0	-1
$-\frac{3\pi}{2}$	$(0, 1)$	0	1
2π	$(1, 0)$	1	0

The range of Q is the set of ordered pairs of numbers $\{(x, y): x^2 + y^2 = 1\}$. But the range of the cosine function, like that of the sine function, is just the set of numbers $\{y: -1 \leqq y \leqq 1\}$. Figure 8-5 shows the graphs of $f(t) = \cos t$ and $g(t) = \sin t$.

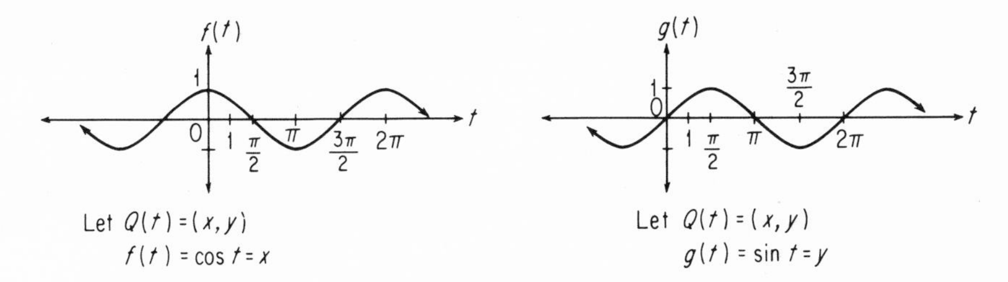

Figure 8-5

With these definitions we can now write

$$Q(t) = (x, y) = (\cos t, \sin t).$$

Since the range of Q is

$$\{(x, y): x^2 + y^2 = 1\}$$

we obtain the important relation

$$(\cos t)^2 + (\sin t)^2 = 1.$$

This identity is usually written in the form

$$\cos^2 t + \sin^2 t = 1.$$

The basic sine and cosine functions have several properties in common:

(1) The domain is the set of real numbers.
(2) The range is $\{y: -1 \leqq y \leqq 1\}$.
(3) The fundamental period is 2π.

Also, the reader will observe that the graphs of the sine and cosine functions appear to be congruent graphs. This is in fact the case. We shall later show that

$$\cos\left(t - \frac{\pi}{2}\right) = \sin t$$

for all numbers t. This means (intuitively) that if the graph of the cosine function could be "moved" (with a rigid motion) $\pi/2$ units in the direction of the positive t-axis, it would become the graph of the sine function.

For each real number t and each integer n,

$$\cos(t + n \cdot 2\pi) = \cos t$$

$$\sin(t + n \cdot 2\pi) = \sin t.$$

Two other important properties of the sine and cosine functions are consequences of the unit circle's symmetry with respect to the x-axis

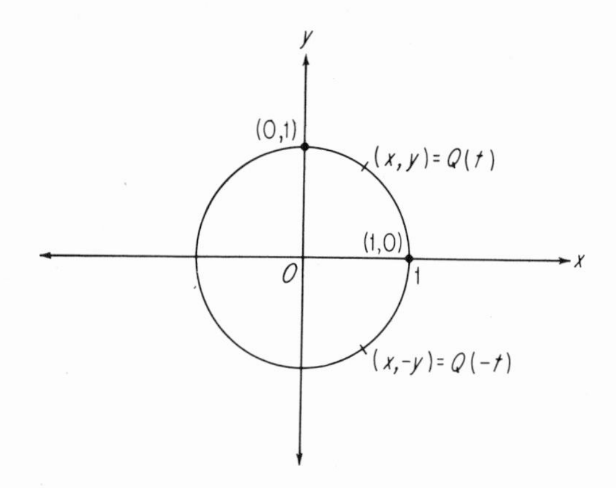

Figure 8-6

The functional values $Q(t)$ and $Q(-t)$ are the coordinates of two points that are symmetric with respect to the x-axis (Figure 8-6).

$$Q(t) = (x, y) = (\cos t, \sin t)$$

$$Q(-t) = (x, -y) = [\cos(-t), \sin(-t)]$$

From this we obtain two generalizations: For each real number t,

$$\cos(-t) = \cos t$$

and

$$\sin(-t) = -\sin t.$$

When we were working with the square functions of Section 8-2 it was easy to calculate $P(t)$, $C(t)$, and $S(t)$ for any real t. But the situation is different for the corresponding circular functions. It is only when t is specially chosen that the calculation of $\cos t$ and $\sin t$ can be easily made. What is more, even though t is a rational number, $\cos t$ and $\sin t$ may be irrational numbers; conversely, for some irrational numbers t, $\cos t$ and $\sin t$ are rational numbers. As the reader might expect, tables of functional values for the sine and cosine functions have been worked out. Table 2 (Appendix) is such a table. With it we can approximate $\cos t$ and $\sin t$ for every real number t. The use of the table will be explained in Section 8-5.

Exact values of $\cos t$ and $\sin t$ for such special values of t as $\pi/6$, $\pi/4$, and $\pi/3$ can be determined without Table 2. All that is needed are some facts from the elementary geometry of circles and triangles. The following table lists some of these special functional values. The student is asked to prove that these are correct as an exercise.

t	$\cos t$	$\sin t$
$\frac{\pi}{6}$	$\frac{\sqrt{3}}{2}$	$\frac{1}{2}$
$\frac{\pi}{4}$	$\frac{\sqrt{2}}{2}$	$\frac{\sqrt{2}}{2}$
$\frac{\pi}{3}$	$\frac{1}{2}$	$\frac{\sqrt{3}}{2}$

8-3 EXERCISES

1. Justify the statement that the range of the cosine function and the range of the sine function are both $\{y: -1 \leqq y \leqq 1\}$.

2. (a) Find $\cos t$, given that $\cos t > 0$ and $\sin t = \frac{3}{5}$.
(b) Find $\sin p$, given that $\sin p < 0$ and $\cos p = \frac{5}{13}$.

3. Use the basic definitions of the circular functions to complete the following table of functional values:

t	$Q(t)$	$\cos t$	$\sin t$
$\frac{3\pi}{2}$			
$-\frac{3\pi}{2}$			
5π			
-5π			
6π			
-6π			
$\frac{7\pi}{2}$			
$-\frac{7\pi}{2}$			
8π			
-8π			
1000π			

4. Use elementary geometry (as well as the basic definitions of the circular functions) to prove that these functional values are correct.

t	$\cos t$	$\sin t$
$\frac{\pi}{6}$	$\frac{\sqrt{3}}{2}$	$\frac{1}{2}$
$\frac{\pi}{4}$	$\frac{\sqrt{2}}{2}$	$\frac{\sqrt{2}}{2}$
$\frac{\pi}{3}$	$\frac{1}{2}$	$\frac{\sqrt{3}}{2}$
$\frac{2\pi}{3}$	$-\frac{1}{2}$	$\frac{\sqrt{3}}{2}$
$\frac{5\pi}{6}$	$-\frac{\sqrt{3}}{2}$	$\frac{1}{2}$

5. Given $\cos(\pi/4) = \sqrt{2}/2$ and $\sin(\pi/4) = \sqrt{2}/2$, use the definitions and the properties of the circular functions and use the geometric symmetries of the unit circle to complete the following table of functional values:

t	$\cos t$	$\sin t$
$\frac{3\pi}{4}$		
$\frac{5\pi}{4}$		
$\frac{7\pi}{4}$		
$\frac{37\pi}{4}$		
$-\frac{\pi}{4}$		
$-\frac{7\pi}{4}$		
$-\frac{9\pi}{4}$		

6. Complete this table of functional values:

t	$\cos t$	$\sin t$
$\frac{4\pi}{3}$		
$-\frac{\pi}{6}$		
$\frac{7\pi}{6}$		
$\frac{22\pi}{3}$		
$\frac{7\pi}{3}$		
$\frac{11\pi}{6}$		
$-\frac{11\pi}{6}$		

7. The domain of each of these functions is given to be all real numbers. What is the range of each function?

(a) $y = 2 \sin t$

(b) $y = \sin (2t)$

(c) $y = |\sin t|$

(d) $y = \cos (t - \pi)$

(e) $y = \sin \left(t + \frac{\pi}{2}\right)$

(f) $y = -\sin t$

(g) $y = t + \sin t$

(h) $y = t \cdot \cos t$

8. Each of these functions is closely related to the basic cosine or sine function and each is periodic. Sketch the graphs. State the fundamental period of each function.

(a) $y = 2 \sin t$ (b) $y = |\sin t|$
(c) $y = |\cos t|$ (d) $y = \sin (2t)$
(e) $y = \cos (t - \pi)$ (f) $y = \sin (t + 2\pi)$
(g) $y = -\sin t$ (h) $y = -\cos t$
(i) $y = \sin \left(t + \frac{\pi}{2}\right)$ (j) $y = 2 \sin \frac{t}{2}$

9. Sketch the graphs of these two functions. (Observe that one of the functions is not periodic.)

(a) $y = \sin t + \cos t$ (b) $y = t + \sin t$

[Note: The results of the following exercise will be used in the proof of a theorem of Section 8-4.]

10. If t denotes a real number, then the Q function (defined in this section) maps t onto the coordinates of a point that lies on the unit circle.

(a) Let a and b denote arbitrary real numbers such that $a + b$ is *not* an integral multiple of 2π. There is a chord of the unit circle whose endpoints are designated by $Q(0)$ and $Q(a + b)$. Also, there is a chord of the unit circle whose endpoints are designated by $Q(b)$ and $Q(-a)$. Then it can be shown that these two chords are of equal length. Give a convincing intuitive argument that this is the case. (You may use the theorem from geometry that if two chords of the same circle intercept arcs that are of equal length, then the chords are of equal length.)

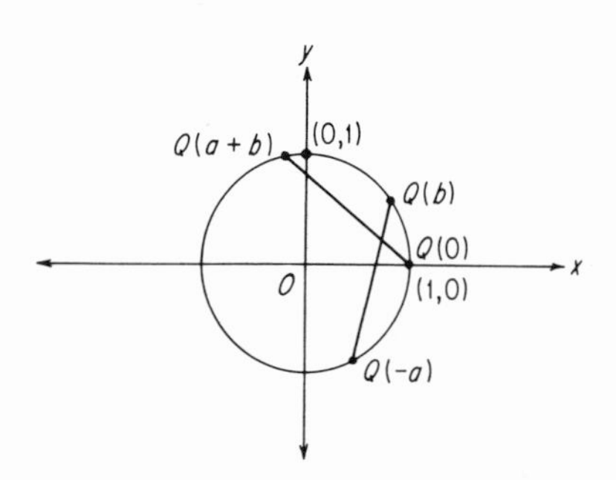

(b) Suppose a and b denote real numbers such that $a + b = 2n\pi$ for some integer n. Then $Q(a + b)$ and $Q(0)$ denote the same point. Show that $Q(b)$ and $Q(-a)$ designate the same point on the unit circle.

8-4 THE ADDITION FORMULAS

Before we explain the use of Table 2 in the Appendix, we need to establish two important facts about the cosine and sine functions:

(1) $\cos(a + b) = \cos a \cos b - \sin a \sin b$.
(2) $\sin(a + b) = \sin a \cos b + \cos a \sin b$.

These facts are known as the *addition formulas*. We shall prove (1) by a geometric argument. Then the student will be asked (in Exercise 2 at the end of this section) to derive the addition formula for $\sin(a + b)$.

Theorem 8-1: For all real numbers a and b,

$$\cos(a + b) = \cos a \cos b - \sin a \sin b.$$

Proof: Suppose that $a + b$ is not an integral multiple of 2π. Let Q denote the Q function of the last section. Consider the chord of the unit circle that has endpoints at $Q(a + b)$ and $Q(0)$ and the chord that has endpoints at $Q(b)$ and $Q(-a)$. By Exercise 10 of Section 8-3 we know that these two chords have the same length (Figure 8-7).

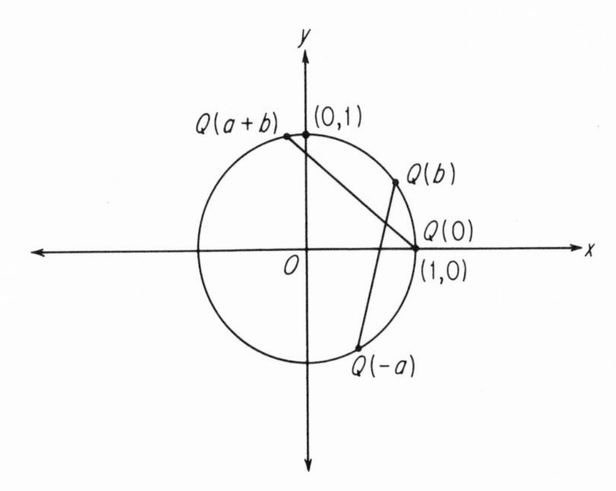

Figure 8-7

From the basic definitions of the cosine and sine we know

$$Q(a + b) = \big(\cos(a + b), \sin(a + b)\big)$$
$$Q(0) = (1, 0)$$
$$Q(b) = (\cos b, \sin b)$$
$$Q(-a) = \big(\cos(-a), \sin(-a)\big) = (\cos a, -\sin a)$$

Since the length of one chord is equal to that of the other chord, then the squares of these lengths will be the same. We shall equate the squares of the lengths of these two chords, using the formula developed in Section 1-11 for calculating the distance between two points.

$$[\cos (a + b) - 1]^2 + \sin^2 (a + b)$$
$$= (\cos b - \cos a)^2 + (\sin b + \sin a)^2$$

Now make use of the identity $\sin^2 x + \cos^2 x = 1$ to establish that

$$-2 \cos (a + b) + 2 = 2 - 2 \cos a \cos b + 2 \sin a \sin b$$

This implies

$$\cos (a + b) = \cos a \cos b - \sin a \sin b$$

The proof is complete for the case in which $a + b$ is not an integral multiple of 2π. If $a + b = 2n\pi$, then $Q(a + b)$ and $Q(0)$ are the same point. Therefore, the distance between $Q(a + b)$ and $Q(0)$ is 0. Also $Q(b)$ and $Q(-a)$ are the same point (Exercise 10, Section 8-3). Therefore, the distance between $Q(b)$ and $Q(-a)$ is 0. The distance formula is valid in the special case when the distance between the two points is 0. Therefore, we can equate the square of the distance between $Q(a + b)$ and $Q(0)$ to the square of the distance between $Q(b)$ and $Q(-a)$. The rest of the proof is identical with that for the case in which $a + b$ is not an integral multiple of 2π.

Two subtraction formulas can be derived from the addition formulas:

(3) $\cos (a - b) = \cos a \cos b + \sin a \sin b$.
(4) $\sin (a - b) = \sin a \cos b - \cos a \sin b$.

The student is asked to derive these formulas in the exercises that follow.

The addition and subtraction formulas are often written together:

$$\cos (a \pm b) = \cos a \cos b \mp \sin a \sin b$$
$$\sin (a \pm b) = \sin a \cos b \pm \cos a \sin b$$

8-4 EXERCISES

1. The reader might be tempted to think that $\cos (a + b) = \cos a + \cos b$. Give a counterexample to show this is *not* a valid generalization.

2. Show that for all numbers a and b:

(a) $\cos(a - b) = \cos a \cos b + \sin a \sin b$.
[Hint: $\cos(a - b) = \cos[a + (-b)]$.]

(b) $\cos\left(a - \frac{\pi}{2}\right) = \sin a$.

(c) $\cos(a - \pi) = -\cos a$.

(d) $\sin\left(a - \frac{\pi}{2}\right) = -\cos a$.

[Hint: $\sin\left(a - \frac{\pi}{2}\right) = \cos(a - \pi)$ by part (b).]

(e) $\sin(a + b) = \sin a \cos b + \cos a \sin b$.

[Hint: $\sin(a + b) = \cos\left[(a + b) - \frac{\pi}{2}\right] = \cos\left[a + \left(b - \frac{\pi}{2}\right)\right]$

by part (b).]

(f) $\sin(a - b) = \sin a \cos b - \cos a \sin b$.

3. Use the addition and subtraction formulas to derive these results:

(a) $\sin\left(\theta + \frac{\pi}{2}\right) = \cos\theta$

(b) $\cos\left(\theta + \frac{\pi}{2}\right) = -\sin\theta$

(c) $\sin\left(\frac{\pi}{2} - \theta\right) = \cos\theta$

(d) $\cos\left(\frac{\pi}{2} - \theta\right) = \sin\theta$

4. Each of these expressions can be reduced to $\pm \sin\theta$ or $\pm \cos\theta$. Make this reduction, using the periodicity of the functions and using the addition and subtraction formulas.

(a) $\sin(2\pi + \theta)$

(b) $\cos(2\pi + \theta)$

(c) $\sin(2\pi - \theta)$

(d) $\cos(2\pi - \theta)$

(e) $\sin\left(\frac{5\pi}{2} - \theta\right)$

(f) $\cos\left(\frac{5\pi}{2} + \theta\right)$

(g) $\sin(43\pi + \theta)$

(h) $\cos(43\pi + \theta)$

(i) $\sin\left(\frac{17\pi}{2} + \theta\right)$

(j) $\cos\left(\frac{17\pi}{2} + \theta\right)$

(k) $\sin\left(-\frac{23\pi}{2} + \theta\right)$

(l) $\cos\left(-\frac{23\pi}{2} + \theta\right)$

5. Show that for all x, $\sin(2x) = 2\sin x \cos x$. [Hint: Use the formula for $\sin(a + b)$.]

6. Prove that these are valid generalizations about $\cos 2x$.

(a) $\cos 2x = \cos^2 x - \sin^2 x$.

(b) $\cos 2x = 2\cos^2 x - 1$.

(c) $\cos 2x = 1 - 2\sin^2 x$.

7. (a) Show by the addition formulas that $\cos 3x = 4\cos^3 x - 3\cos x$. [Hint: $\cos 3x = \cos(2x + x)$.]
   (b) Derive a similar expression for $\sin 3x$.

8-5 MORE ON THE SINE AND COSINE FUNCTIONS

Table 2 in the Appendix can be used to approximate $\sin x$ and $\cos x$ for any real number x. It is easy to use if the number x is on the interval $0 \leqq x \leqq \pi/2$.

For example,

$$\sin 0.1 \approx 0.0998$$
$$\sin 0.58 \approx 0.5480$$
$$\cos 1.56 \approx 0.0108.$$

If x is on the interval $0 \leqq x \leqq \pi/2$ but is not one of the specific entries in the table, then the use of linear interpolation (Section 3-6) is appropriate to approximate $\sin x$ and $\cos x$.

$$\begin{aligned}\sin 1.316 &\approx \sin 1.31 + \frac{\sin 1.32 - \sin 1.31}{1.32 - 1.31}(1.316 - 1.31)\\ &\approx 0.9662 + \frac{0.9687 - 0.9662}{0.01}(0.006)\\ &\approx 0.9677\end{aligned}$$

The reader will observe that there are only a few entries in Table 2 for x outside the interval $0 \leqq x \leqq \pi/2$. How can the table be used to approximate $\sin 2.00$ or $\cos 3.10$? The answer is: Use the addition formulas and the fact that $\pi/2 \approx 1.57$.

$$\begin{aligned}\sin 2.00 &\approx \sin\left(\frac{\pi}{2} + 0.43\right)\\ &= \cos 0.43\\ &\approx 0.9090\end{aligned}$$

$$\begin{aligned}\cos 3.10 &\approx \cos\left(\frac{\pi}{2} + 1.53\right)\\ &= -\sin 1.53\\ &\approx -0.9992\end{aligned}$$

Every number x can be written in the form

$$n\frac{\pi}{2} + r$$

where n is an integer and r is a number on the interval

$$0 \leqq r < \frac{\pi}{2}.$$

Therefore,

$$\sin x = \sin\left(n\frac{\pi}{2} + r\right)$$

$$\cos x = \cos\left(n\frac{\pi}{2} + r\right).$$

The use of an addition formula will reduce either of these to $\pm \sin r$ or $\pm \cos r$, where r is a number between 0 and $\pi/2$.

Thus a relatively short table of values suffices to approximate $\sin x$ and $\cos x$ for every possible choice of x. The reader should be reminded also that $\sin(-x) = -\sin x$ and $\cos(-x) = \cos x$.

Here are four more examples of the use of Table 2.

$$\begin{aligned}
\sin 2.54 &\approx \sin\left(\frac{\pi}{2} + 0.97\right) \\
&= \cos 0.97 \\
&\approx 0.5653 \\
\cos(-4.83) &= \cos 4.83 \\
&\approx \cos\left(\frac{3\pi}{2} + 0.12\right) \\
&= \sin 0.12 \\
&\approx 0.1197 \\
\sin 100 &\approx \sin\left(\frac{63\pi}{2} + 1.09\right) \\
&= -\cos 1.09 \\
&\approx -0.4625 \\
\cos 4.005 &\approx \cos(\pi + 0.863) \\
&= -\cos 0.863 \\
&\approx -0.6501
\end{aligned}$$

Linear interpolation was used in the last example.

8-5 EXERCISES

1. Approximate each of the following numbers by a number in the form $n(\pi/2) + r$, where n is an integer and $0 \leqq r < \pi/2$. Round off r to two decimal places ($\pi/2 \approx 1.571$).

(a) 2.45 (b) 5.74
(c) 30 (d) 7.16
(e) 1.8 (f) 4.73
(g) 3 (h) 3π

2. From Table 2 find an approximation for each of the following:

(a) $\cos 2.45$ (b) $\cos 6.28$
(c) $\sin 1.8$ (d) $\sin 12$
(e) $\sin(-4.73)$ (f) $\cos(-1.34)$

3. Use Table 2 and linear interpolation (if needed) to find an approximation for each of the following:

(a) $\cos 1.550$ (b) $\sin 0.392$
(c) $\cos 0.372$ (d) $\sin(-1.114)$
(e) $\cos(-1.377)$ (f) $\sin(0.976)$

8-6 FUNCTIONS DEFINED BY $f(x) = A \sin(Bx + C)$

Let A and B denote positive numbers and let C denote any number. The rule $f(x) = A \sin(Bx + C)$ can be thought of as defining a three-parameter family of periodic functions. (The domain of each function is the set of all real numbers unless otherwise specified.) The members of this family deserve special attention, because they have so many applications in physics.

The function F defined by $F(x) = \sin x$ is a member of the family—the member obtained by taking $A = 1$, $B = 1$, and $C = 0$. We shall use this function as a benchmark (Figure 8-8); other.

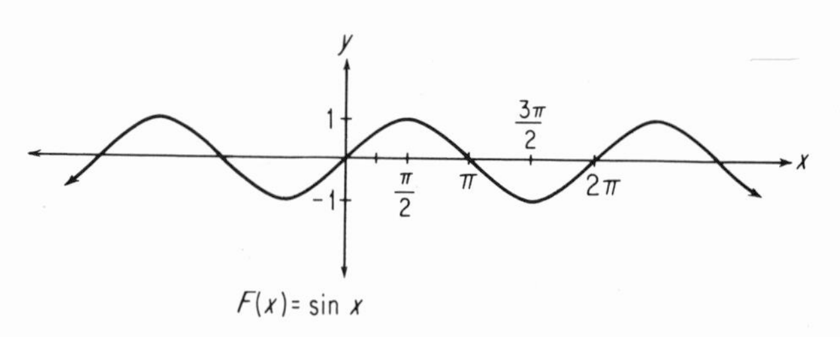

Figure 8-8

members of the family will be compared to it. The reader will recall that the range of F is $\{y: -1 \leqq y \leqq 1\}$ and the fundamental period is 2π. Physicists sometimes refer to the graph as a "sine wave." The "wave length" (period) is 2π and the "amplitude" is 1. In general, if the range of a periodic function contains a maximum number M and a minimum number m, then the **amplitude** of the function is defined to be $\frac{1}{2}|M - m|$.

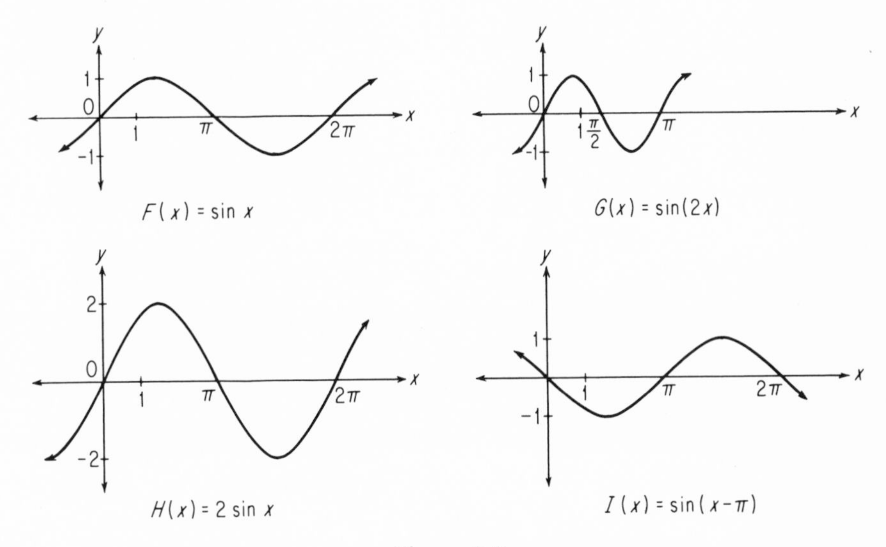

Figure 8-9

We now compare the graphs (Figure 8-9) of three other members of the family to the graph of F. Let

$$G(x) = \sin (2x)$$
$$H(x) = 2 \sin x$$
$$I(x) = \sin (x - \pi).$$

The function G has the same range as F, but its fundamental period (wave length) is π. This can be seen by the following reasoning: As x varies from 0 to π, $2x$ varies from 0 to 2π. Therefore, $\sin (2x)$ completes a cycle on the interval from 0 to π. The wave length of G is only half that of F.

Since $H(x) = 2 \sin x$, the rule for H also could be written $H(x) = 2F(x)$. The function H has the same wave length as F, but a different range: $\{y: -2 \leqq y \leqq 2\}$. The amplitude of H is twice the amplitude of F.

Finally, the graph of I is congruent to the graph of F—but it is shifted π units to the right. (See Section 6-3, Exercise 13.) Physicists sometimes refer to this as a "phase shift." The functions I and F have the same range and the same fundamental period, but they are not in phase.

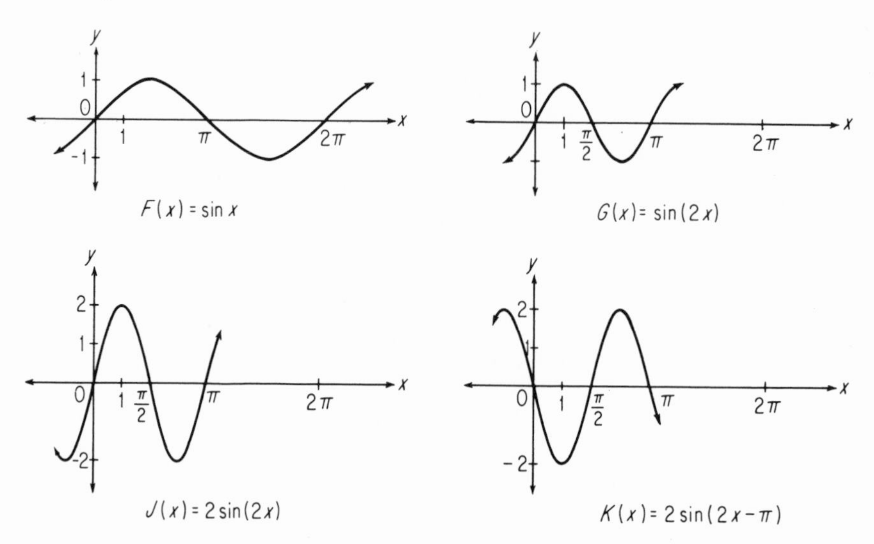

Figure 8-10

Now consider the sequence of functions (Figure 8-10):

$$F(x) = \sin x$$

$$G(x) = \sin (2x)$$

$$J(x) = 2 \sin (2x)$$

$$K(x) = 2 \sin (2x - \pi)$$

The functions F and G have already been compared. The rule for J can be written $J(x) = 2G(x)$, so J and G have the same fundamental period, but different ranges.

The graph of K, the last function in the sequence of four, is congruent to that of J, but the two are not in phase:

$$K(x) = 2 \sin 2\left(x - \frac{\pi}{2}\right) = J\left(x - \frac{\pi}{2}\right)$$

The graph of J can be translated $\pi/2$ units to the right to obtain the graph of K.

In order to make general statements about the members of the family $f(x) = A \sin (Bx + C)$, consider the sequence of functions:

$$F(x) = \sin x$$

$$G(x) = \sin Bx$$

$$H(x) = A \sin Bx$$

$$I(x) = A \sin (Bx + C)$$

In this sequence A and B denote arbitrary positive numbers, and C denotes any number. We shall compare G to F, H to G, and I to H.

The functions G and F have the same range, but the fundamental period of G is $2\pi/B$. (As x varies from 0 to $2\pi/B$, Bx varies from 0 to 2π.)

The functions H and G have the same fundamental period $2\pi/B$, but the range of H is $\{y: -A \leqq y \leqq A\}$.

The functions I and H have the same fundamental period and the same range; their graphs are congruent, but not in phase:

$$I(x) = A \sin B\left(x + \frac{C}{B}\right) = H\left(x + \frac{C}{B}\right)$$

Translate the graph of H by C/B units to obtain the graph of I. The translation is to the left if C is positive, to the right if C is negative.

As a final example, consider the function (Figure 8-11) de-

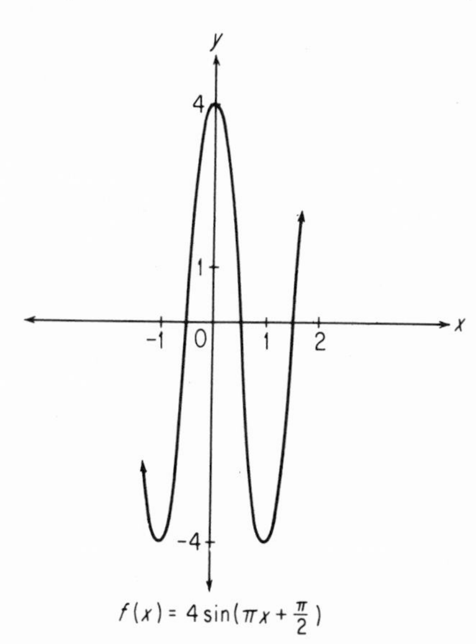

Figure 8-11

fined by $f(x) = 4 \sin (\pi x + \pi/2)$. The range of this function is $\{y: -4 \leqq y \leqq 4\}$; its fundamental period is 2. The graph is $\frac{1}{2}$ unit out of phase with the graph of $g(x) = 4 \sin \pi x$.

8-6 EXERCISES

1. Sketch one full cycle of the graph of each of these functions:

(a) $y = \sin \dfrac{x}{2}$ (b) $y = 4 \sin \dfrac{x}{2}$

(c) $y = 2 \sin \left(x - \dfrac{3\pi}{2}\right)$ (d) $y = \sin 2\pi x$

(e) $y = \sin (2\pi x + \pi)$ (f) $y = 2 \sin (2\pi x + \pi)$

(g) $y = \frac{1}{2} \sin (\frac{1}{2}x)$ (h) $y = \frac{1}{2} \sin (\frac{1}{2}x + \pi)$

2. Below are the graphs of two functions that belong to the family defined by $f(x) = A \sin (Bx + C)$. Write the specific rule for each function.

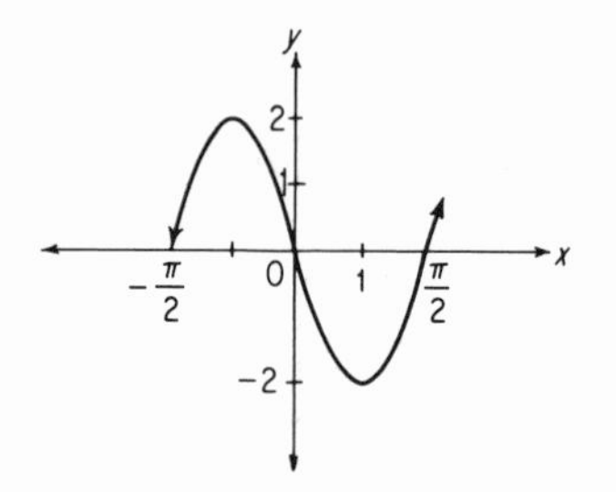

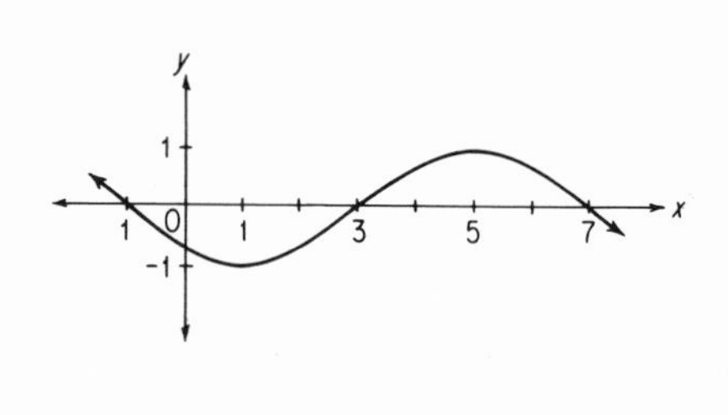

3. By identifying the corresponding values of A, B, and C show that each of these two functions belongs to the family defined by $f(x) = A \sin (Bx + C)$:

(a) $y = \cos x$ (b) $y = 2 \cos (x + \pi)$

4. Let D and E denote positive numbers and F denote any number. The rule $g(x) = D \cos (Ex + F)$ can be thought of as defining a three-parameter family of periodic functions. Show that each member of this family is also a member of the family defined by $f(x) = A \sin (Bx + C)$, where A and B are positive numbers and C is any number. [Hint: Show that A, B, and C can be found (expressed in terms of D, E, and F) such that $D \cos (Ex + F) = A \sin (Bx + C)$.]

†**5.** Let

$$f(x) = \begin{cases} 0, & \text{if } x < 0 \\ \dfrac{1}{x+1} \cdot \cos x, & \text{if } x \geqq 0 \end{cases}$$

This function might be used to describe a "damped oscillation" in physics. Sketch the graph of f. [Hint: First sketch the graph of $g(x) = 1/(x + 1)$ and of its negative.]

8-7 FUNCTIONS CONSTRUCTED FROM THE SINE AND COSINE FUNCTIONS

Suppose that f and g are functions with the same domain D, and suppose that the range of each is a subset of the real numbers. Then f/g is a function that has as its domain all the elements of D for which $g(x) \neq 0$. The rule for f/g is

$$\frac{f}{g}(x) = \frac{f(x)}{g(x)}$$

This is one of the ways of constructing a function from known functions f and g. (See Section 6-1.)

The set of real numbers is the domain for both the sine and cosine functions. We now define four additional functions in terms of the sine or cosine function.

Definition: For all real numbers $x \neq (\pi/2) + n\pi$, where n is an integer,

$$\text{tangent } x = \frac{\sin x}{\cos x}.$$

"Tangent x" is abbreviated as "tan x."

Definition: For all real numbers $x \neq n\pi$, where n is an integer,

$$\text{cotangent } x = \frac{\cos x}{\sin x}.$$

"Cotangent x" is abbreviated as "cot x."

Definition: For all real numbers $x \neq (\pi/2) + n\pi$, where n is an integer,

$$\text{secant } x = \frac{1}{\cos x}.$$

"Secant x" is abbreviated as "sec x."

Definition: For all real numbers $x \neq n\pi$, where n is an integer,

$$\text{cosecant } x = \frac{1}{\sin x}.$$

"Cosecant x" is abbreviated as "csc x."

The tangent and cotangent functions are periodic functions with a fundamental period of π. The range of each is the set of all real numbers. Figure 8-12 shows the graphs of the tangent and cotangent functions over one period.

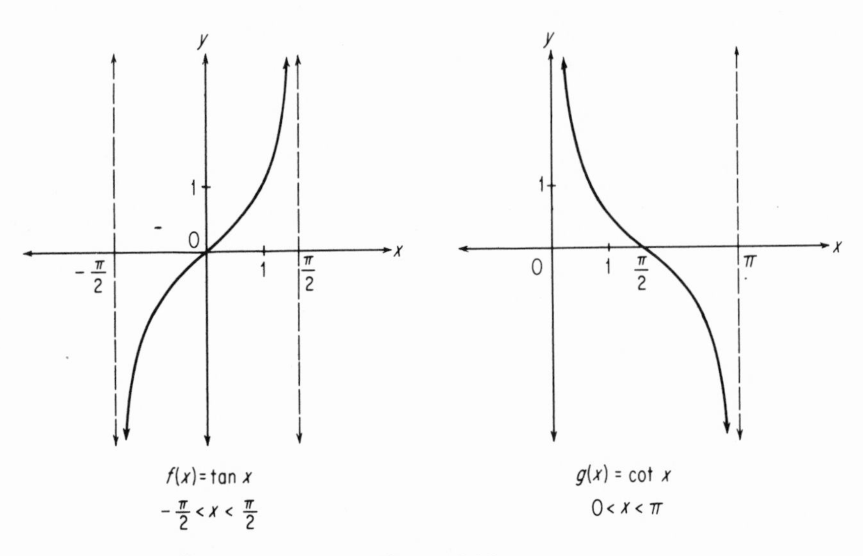

Figure 8-12

The secant and cosecant functions are periodic functions with a fundamental period of 2π. The range of each is $\{y: |y| \geqq 1\}$. Figure 8-13 shows the graphs of the secant and cosecant functions over one period.

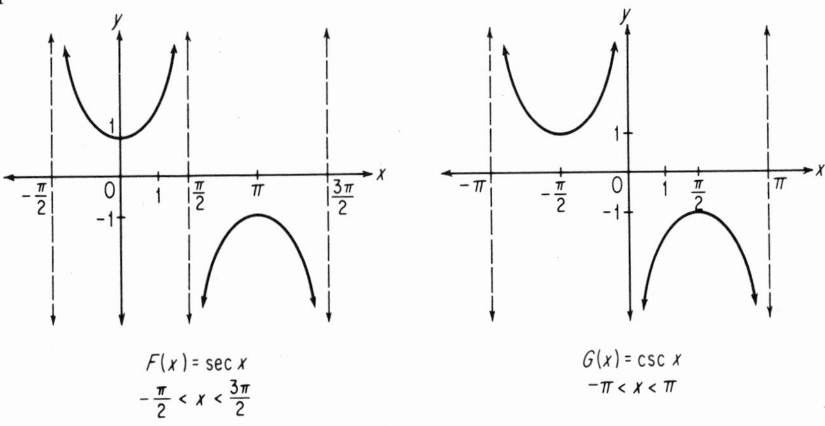

Figure 8-13

Functional values of $\tan x$, $0 \leqq x < \pi/2$, are found in Table 2 of the Appendix. Functional values of $\cot x$, $\sec x$, and $\csc x$ can be obtained by using the reciprocal relationships:

$$\cot x = \frac{1}{\tan x}$$

$$\sec x = \frac{1}{\cos x}$$

$$\csc x = \frac{1}{\sin x}.$$

8-7 EXERCISES

1. Use the definition of the function to find each functional value.

(a) $\sec \frac{\pi}{3} = ?$ (b) $\sec \left(-\frac{\pi}{3}\right) = ?$

(c) $\cot \frac{3\pi}{2} = ?$ (d) $\tan \frac{\pi}{4} = ?$

(e) $\sec 6\pi = ?$ (f) $\cot 5\pi = ?$

(g) $\tan \frac{\pi}{3} = ?$ (h) $\tan \left(-\frac{\pi}{3}\right) = ?$

(i) $\csc \left(-\frac{\pi}{6}\right) = ?$ (j) $\tan \frac{3\pi}{4} = ?$

(k) $\sec \frac{\pi}{6} = ?$ (l) $\sec \frac{\pi}{4} = ?$

(m) $\sec \frac{2\pi}{3} = ?$ (n) $\cot \frac{\pi}{4} = ?$

(o) $\csc \frac{3\pi}{2} = ?$ (p) $\tan 5\pi = ?$

(q) $\cot \frac{\pi}{3} = ?$ (r) $\sec (-105\pi) = ?$

2. Find the zeros of:

(a) $f(x) = \tan x, -\frac{\pi}{2} < x < \frac{\pi}{2}.$

(b) $g(x) = \cot x, 0 < x < \pi.$

(c) $F(x) = \sec x, -\frac{\pi}{2} < x < \frac{3\pi}{2}.$

(d) $G(x) = \csc x, -\pi < x < \pi.$

3. Prove $\tan(-x) = -\tan x$, $x \neq \frac{\pi}{2} + n\pi$, where n is an integer.

4. Prove $\cot(-x) = -\cot x$, $x \neq n\pi$, where n is an integer.

5. Prove $\csc(-x) = -\csc x$, $x \neq n\pi$, where n is an integer.

6. Prove $\sec(-x) = \sec x$, $x \neq \frac{\pi}{2} + n\pi$, where n is an integer.

7. Sketch the graph of each of the following over one period.
(a) $y = \tan(-x)$ (b) $y = \tan 2x$
(c) $y = \cot(-x)$ (d) $y = \cot 2x$
(e) $y = \tan\left(x + \frac{\pi}{2}\right)$ (f) $y = \sec 2x$

8. Prove $\tan\left(\frac{\pi}{2} - t\right) = \cot t$.

9. Prove $\tan(x + y) = \dfrac{\tan x + \tan y}{1 - \tan x \tan y}$.

10. Prove $\tan 2x = \dfrac{2 \tan x}{1 - \tan^2 x}$.

11. Prove $\tan^2 x + 1 = \sec^2 x$. [Hint: $\sin^2 x + \cos^2 x = 1$.]

12. Prove $1 + \cot^2 x = \csc^2 x$.

13. Prove that the tangent function is periodic and has a period of π.

14. Prove that the secant function is periodic with a period of 2π.

8-8 DOES f MAP ANY NUMBER ONTO c?

Suppose that f denotes one of the circular functions, with the domain of f being all real numbers. Does f map any number onto c? If so, what number(s)? These are familiar questions—questions that we have asked previously about linear and quadratic and exponential functions.

In answering such questions the reader should remember the following:

(1) f maps at least one number onto c if and only if c belongs to the range of f.

(2) If f maps x onto c, then x is a root of the equation $f(x) = c$.

(3) The circular functions are periodic; therefore, if f has a fundamental period of p and f maps x onto c, then f also maps $x + np$ (where n is an arbitrary integer) onto c.

EXAMPLE 1: Let the domain of f be the real numbers and let the rule be $f(x) = \cos x$ (Figure 8-14). We shall pose a series of

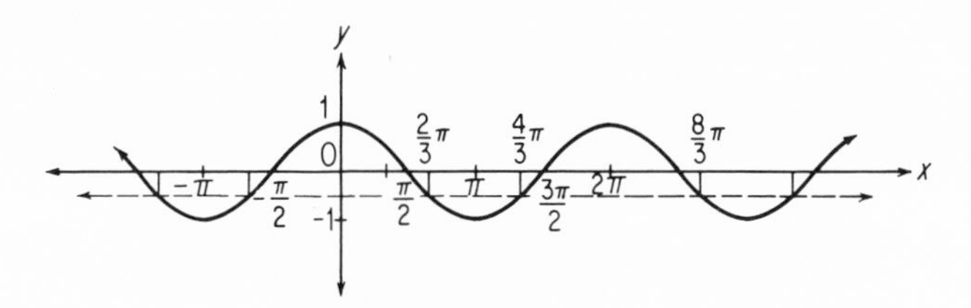

Figure 8-14

questions (Q) about the cosine function and after each question supply the answer (A).

Q: Does f map any number onto 2?
A: No, 2 is not in the range of f.

Q: Does f map any number onto 0?
A: Yes, $\pi/2$ is mapped onto 0, and there are thousands of others also. In fact, every number of the form $\pi/2 + n\pi$, where n is an integer, will be mapped onto 0.

Q: Does f map any number onto $-\frac{1}{2}$?
A: Yes, $2\pi/3$ is mapped onto $-\frac{1}{2}$. But there are others also. In fact, all numbers of the form $(2n + 1)\pi \pm (\pi/3)$ are mapped onto $-\frac{1}{2}$.

Plainly, the answer to the question, "Does f map any number onto c?" can be complicated. We will avoid such complicated answers by making the rest of our examples pertain to functions that have restricted domains.

EXAMPLE 2: Let F denote the function with domain: $\{x: 0 \leqq x \leqq \pi\}$ and with rule $F(x) = \cos x$. Of course, F is the restriction of f (the function of Example 1) to the closed interval from 0 to π. Notice that the two functions have the same range, even though they have different domains (Figure 8-15). The function F has another property that is important and that makes it easy to work with: it is one-to-one. Any equation of the form $F(x) = c$ will have exactly one solution (root) or none at all.

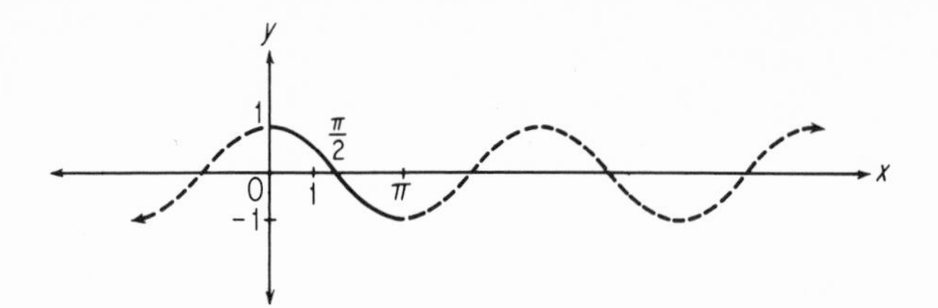

Figure 8-15

Q: Does F map any number onto 0?

A: Yes, $\pi/2$ is mapped onto 0, but nothing else in the domain of this function is mapped onto 0.

Q: Does F map any number onto $-\frac{1}{2}$?

A: Yes, $2\pi/3$ is the only number in the domain that is mapped onto $-\frac{1}{2}$.

Q: Does F map any number onto 0.2?

A: Yes, according to Table 2 there is a number between 1.36 and 1.37 that is mapped onto 0.2. We can use linear interpolation to approximate this number as 1.369.

EXAMPLE 3: Let H have domain: $\{x: -\pi/2 \leqq x \leqq \pi/2\}$ and rule $H(x) = \sin x$. Thus H is the restriction of the basic sine function to the closed interval from $-\pi/2$ to $\pi/2$. The function H is a one-to-one function; its range is $\{y: -1 \leqq y \leqq 1\}$. (See Figure 8-16.)

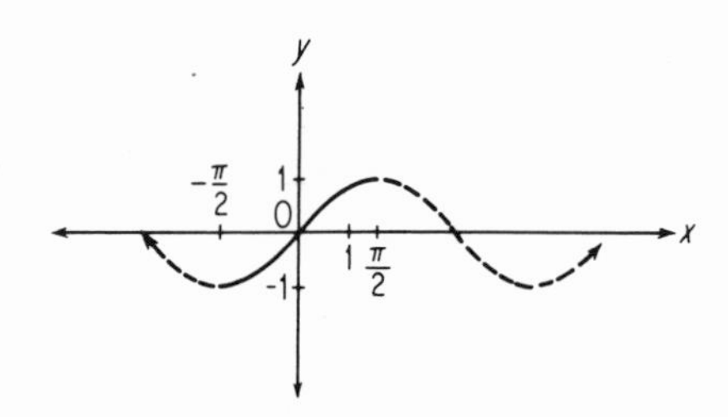

Figure 8-16

Q: Does H map any number onto -1.5?

A: No, -1.5 is not in the range of H.

Q: Does H map any number onto 1?

A: Yes, $\pi/2$ is mapped onto 1.

Q: Does H map any number onto 0.9?

A: Yes, Table 2 shows that $H(1.12) = 0.9001$. So the number that is mapped onto 0.9 is slightly less than 1.12.

Q: Does H map any number onto -0.9?

A: Yes, -0.9 is certainly in the range of H. Furthermore,

$$\sin x = -0.9 \Leftrightarrow -\sin x = 0.9$$
$$\Leftrightarrow \sin(-x) = 0.9.$$

Therefore, by the preceding answer, $-x \approx 1.12$ and $x \approx -1.12$.

EXAMPLE 4: Let the domain of J be $\{x: -\pi/2 < x < \pi/2\}$ and let the rule be $J(x) = \tan x$. Thus J is the restriction of the basic tangent function to the open interval from $-\pi/2$ to $\pi/2$. The function J is a one-to-one function, and its range is the set of real numbers (Figure 8-17).

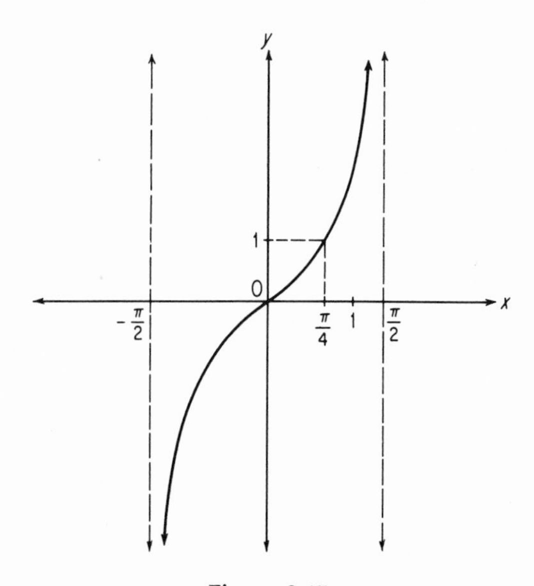

Figure 8-17

Q: Does J map any number onto 0?

A: Yes, 0 is mapped onto 0.

Q: Does J map any number onto 1?

A: Yes, $\pi/4$ is mapped onto 1.

Q: Does J map any number onto 5?

A: Yes, by Table 2 we see that there is a number between 1.37 and 1.38 that is mapped onto 5. Linear interpolation yields 1.373 as our best approximation from this table.

Q: Does J map any number onto -5?

A: Yes, $\tan(-x) = -\tan x$. Therefore, from the preceding answer we know that the number that is mapped onto -5 is approximately -1.373.

Q: Does J map any number onto 10^9?

A: Yes, there is such a number. From Table 2 all we can tell is that it lies between 1.57 and $\pi/2$.

Q: If I give you *any* number c (no matter how large or how small) will there be an affirmative answer to the question, "Does J map some unique number onto c?"

A: Yes, because the range of J is all real numbers, and J is a one-to-one function.

8-8 EXERCISES

For many of these exercises you will not need to use a table of functional values. For others use Table 2 to approximate the answers.

1. Let $F(x) = \cos x$, domain: $\{x: 0 \leqq x \leqq \pi\}$. Find all the numbers (if any) that F maps onto:

(a) $-\frac{1}{2}$
(b) 1.0101
(c) 1
(d) 0.8870
(e) $\frac{\sqrt{2}}{2}$
(f) $\frac{\sqrt{3}}{2}$
(g) 0
(h) -0.5817

2. Let $H(x) = \sin x$, domain: $\{x: -\pi/2 \leqq x \leqq \pi/2\}$. Find all the numbers (if any) that H maps onto:

(a) $\frac{1}{2}$
(b) 0.5354
(c) -1
(d) $\frac{1}{3}$
(e) $\frac{\sqrt{3}}{2}$
(f) 0.6052
(g) 0
(h) -0.5354

3. Let $J(x) = \tan x$, domain: $\{x: -\pi/2 < x < \pi/2\}$. Find all the numbers (if any) that J maps onto:

(a) -1 (b) 2.5000
(c) $\sqrt{3}$ (d) 0.4346
(e) $\frac{1}{2}$ (f) -1.2600
(g) 1 (h) 0.3333

4. Find all the numbers on the interval $0 \leqq x < 2\pi$ such that:

(a) $2 \sin x = 1$
(b) $\tan^2 x = 1$
(c) $\cos^2 x = 1$
(d) $\sin x = \cos x$
(e) $\sin^2 x = \sin x$ [Hint: This is a quadratic equation in $\sin x$.]

5. Let f be a periodic function with a fundamental period of p. Consider f over an interval whose length is p and suppose x_1 and x_2 are the only elements on this interval that f maps onto c. Then the set of all real solutions of the equation $f(x) = c$ is:

$$\{x: x = x_1 + n \cdot p, n \text{ an integer}\} \cup \{x: x = x_2 + n \cdot p, n \text{ an integer}\}$$

Find the set of all real solutions of:

(a) $\sin x = \frac{1}{2}$ (b) $\tan x = 1$
(c) $\cos x = 1$ (d) $\sin x = 0.1593$

8-9 THE ARCCOSINE, ARCSINE, AND ARCTANGENT FUNCTIONS

The reader will recall that the only functions for which inverse functions can be constructed are those that are one-to-one. The function f, defined by $f(x) = \cos x$, domain: the real numbers, is certainly *not* a one-to-one function, so there is no such thing as the function that is the inverse of f. However, a suitable restriction of the

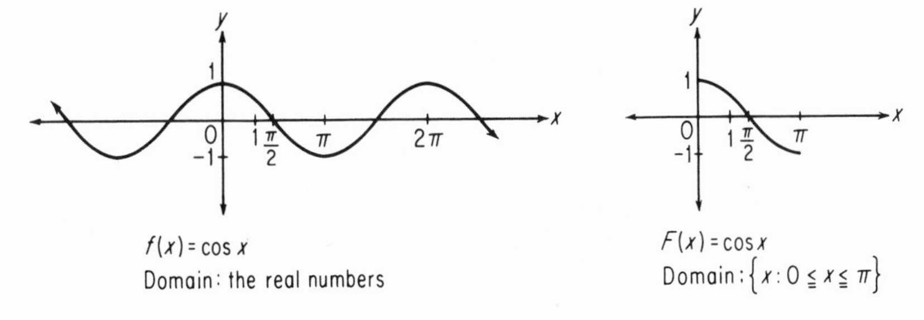

Figure 8-18

domain of the cosine function will produce a function that has the same range as $f(x) = \cos x$ and that *is* one-to-one, and thus has an inverse function.

Let $F(x) = \cos x$, domain: $\{x: 0 \leqq x \leqq \pi\}$. It is apparent from the definition of $\cos x$ that F is decreasing on its domain. Since it is decreasing, F is also a one-to-one function and therefore has an inverse function. We know that the inverse function exists and we know that it is defined as in Section 6-4. It is customary to call this inverse function the Arccosine function (with a capital A) and denote its functional value by the abbreviation Arccos x. This means that $G(x) = \text{Arccos } x$, domain: $\{x: -1 \leqq x \leqq 1\}$ is, by definition, the function that is the inverse of $F(x) = \cos x$, domain: $\{x: 0 \leqq x \leqq \pi\}$.

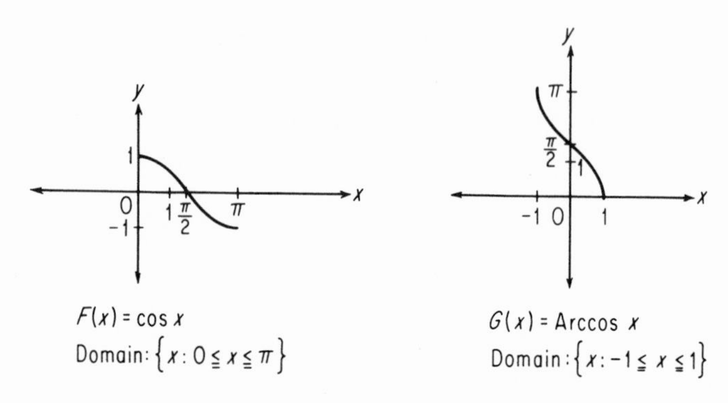

Figure 8-19

The graph of G can be obtained by reflecting the graph of F in the line $y = x$. (See Figure 8-19.) The range of G is the same as the domain of F, $\{y: 0 \leqq y \leqq \pi\}$.

Since $\cos \pi = -1$, it follows that $\text{Arccos}(-1) = \pi$.

Since $\cos 0 = 1$, it follows that $\text{Arccos } 1 = 0$.

Since $\cos \dfrac{\pi}{3} = \dfrac{1}{2}$, it follows that $\text{Arccos } \dfrac{1}{2} = \dfrac{\pi}{3}$.

In a similar way the Arcsine and Arctangent functions are defined as the inverses of suitable restrictions of the sine and tangent functions. The abbreviations Arcsin x and Arctan x are used to denote the functional values.

Let $H(x) = \sin x$, domain: $\{x: -\pi/2 \leqq x \leqq \pi/2\}$. Then the function I that has the set $\{x: -1 \leqq x \leqq 1\}$ as its domain and that has the rule $I(x) = \text{Arcsin } x$ is (by definition) the inverse of H. The

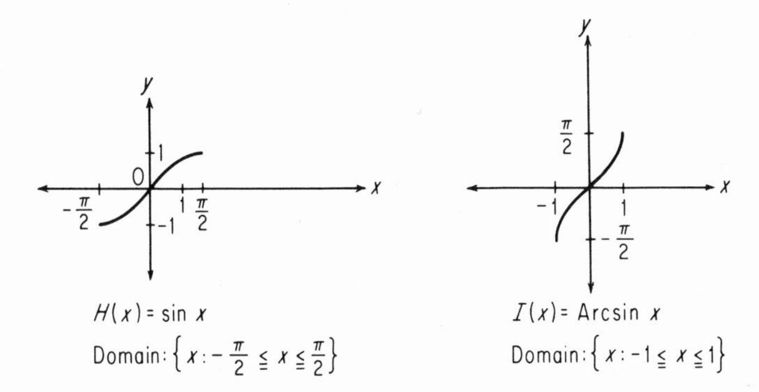

Figure 8-20

range of the Arcsin function is $\{y: -\pi/2 \leqq y \leqq \pi/2\}$. Figure 8-20 shows the graphs of H and I.

Let $J(x) = \tan x$, domain: $\{x: -\pi/2 < x < \pi/2\}$. Then the function K that has the real numbers as its domain and has the rule $K(x) = \text{Arctan } x$ is (by definition) the inverse of J. (See Figure 8-21.)

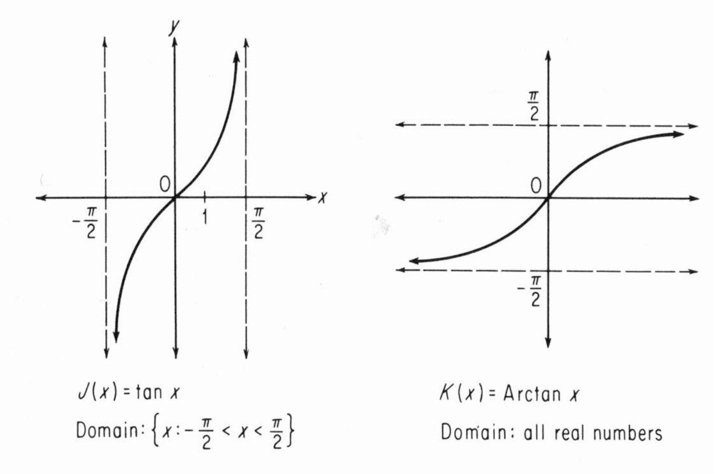

Figure 8-21

The range of the Arctan function is $\{y: -\pi/2 < y < \pi/2\}$.

$$\text{Since } \sin \frac{\pi}{2} = 1, \text{ then } \text{Arcsin } 1 = \frac{\pi}{2}.$$

$$\text{Since } \sin \left(-\frac{\pi}{6}\right) = -\frac{1}{2}, \text{ then } \text{Arcsin} \left(-\frac{1}{2}\right) = -\frac{\pi}{6}.$$

$$\text{Since } \tan \frac{\pi}{4} = 1, \text{ then } \text{Arctan } 1 = \frac{\pi}{4}.$$

$$\text{Since } \tan\frac{\pi}{6} = \frac{\sqrt{3}}{3}, \text{ then } \operatorname{Arctan}\frac{\sqrt{3}}{3} = \frac{\pi}{6}.$$

8-9 EXERCISES

1. Simplify:

(a) Arcsin (-1) (b) Arcsin $\left(\frac{\sqrt{3}}{2}\right)$

(c) Arccos $(-\frac{1}{2})$ (d) Arctan (-1)

(e) Arcsin 0 (f) Arccos 0

(g) Arctan 0 (h) sin [Arccos (-1)]

(i) cos (Arccos $\frac{1}{3}$) (j) tan (Arctan 4)

2. Use a table of values to approximate:

(a) Arcsin 0.8 (b) Arccos (-0.2)

(c) Arctan 40 (d) Arctan (-10)

(e) Arctan (5000) (f) Arcsin (0.01)

(g) Arcsin (-0.8) (h) Arccos (0.01)

3. The symbols Arccos x, Arcsin x, and Arctan x give us short names for numbers that would otherwise have rather long descriptions. For example, "the smallest positive number x such that $\cos x = 0.1$" is now named Arccos (0.1). In a similar way, give brief names to each of these numbers.

(a) The only number x between $-\pi/2$ and $\pi/2$ such that $\sin x = 0.4$.

(b) The only number x between $-\pi/2$ and $\pi/2$ such that $\tan x = 47$.

(c) The only number x between 0 and π such that $\cos x = -0.6$.

(d) The smallest positive number x such that $\tan x = \frac{1}{3}$.

4. (a) Arccos $(\sin \pi/6) = ?$ (b) Arctan $(\cot 2\pi/3) = ?$

(c) Arccos $(\cos 2\pi/3) = ?$ (d) Arcsin $(\sin 2\pi/3) = ?$

†5. An Arccotangent function can be defined in a fashion analogous to the other definitions of this section. Suggest an appropriate definition for an Arccotangent function and then sketch the graph of the function you define.

8-10 HOW TO MEASURE AN ANGLE

The circular functions have many applications to triangles and angles. To understand these applications it is necessary to know what is meant by the *measure* of an angle.

The union of two rays that have a common endpoint is called an **angle.** The common endpoint is known as the **vertex.** When three points are used to designate an angle, the symbol for the vertex is placed in the middle (Figure 8-22).

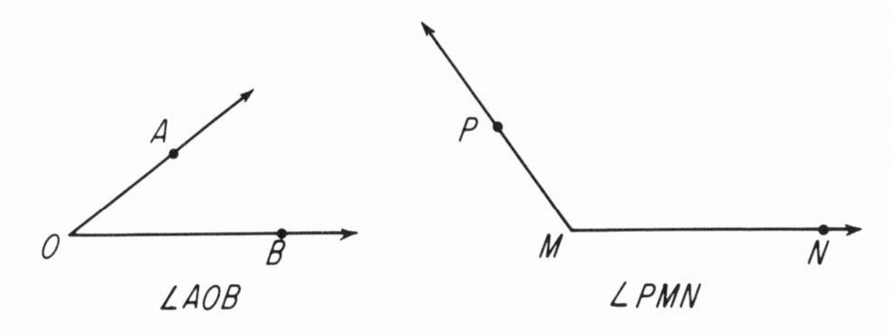

Figure 8-22

As everyone knows, angles come in different "sizes." How is an angle measured and in what units?

Perhaps the first thing that should be said is that there are several commonly used units of angle measurement. Just as the length of a line segment can be given in feet, inches, or centimeters, so can the measure of an angle be given in *radians*, *degrees*, or *mils*. The unit of measurement that we shall emphasize is the *radian*.

Suppose an arbitrary angle has its vertex at point V. Choose a unit length and construct the unit circle that has center at point V (and is in the plane of the angle). The unit circle meets the angle at either a single point or at two points (Figure 8-23).

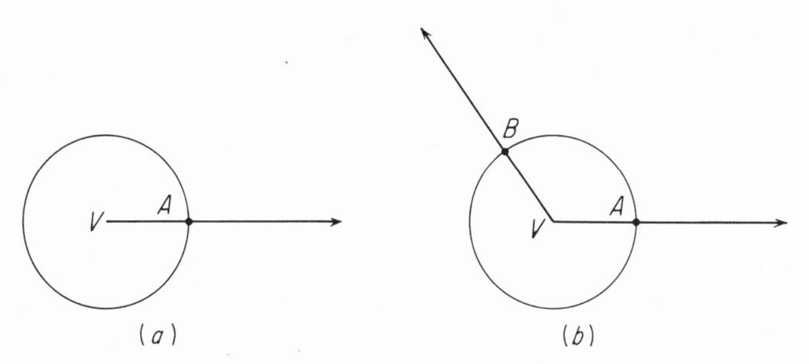

Figure 8-23

If the angle is the union of two collinear rays that have the same direction (Figure 8-23a), then the unit circle meets the angle at a single point A. We define the radian measure of such an angle to be 0.

If the angle is not the union of two collinear rays with the same direction, then the unit circle meets the angle at two points A and B. Let L denote the line (Figure 8-24) that is perpendicular to the ray VA at point A. (It would be just as suitable to let L denote the line that is

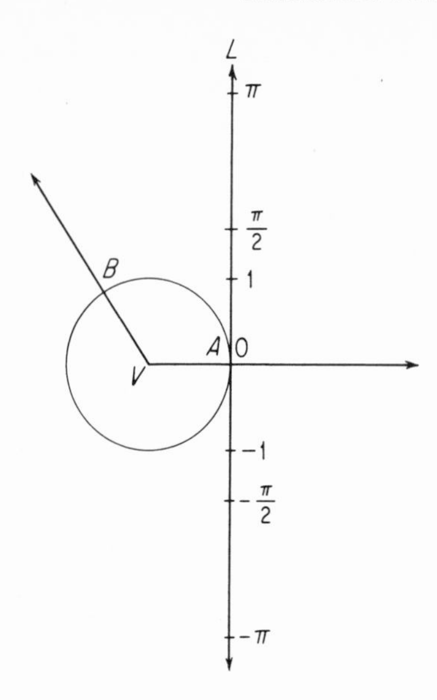

Figure 8-24

perpendicular to the ray VB at point B.) Now let L be "coordinatized" (as in Section 1-1) in such a way that:

(1) The point A is assigned the coordinate 0.
(2) The line segment from the point with coordinate 0 to the point with coordinate 1 has the same length as a radius of the unit circle.

In Section 8-3 we assumed that it is possible to wrap the line L around the unit circle in such a way that each point of L is mapped onto a unique point of the unit circle.

The circumference of the unit circle is 2π. The half-open line segment that is a subset of L and that is the graph of

$$\{x: -\pi < x \leqq \pi\}$$

also has length 2π. We shall assume that if L is wrapped around the unit circle *each point of the circle is the image of exactly one point on the line segment* $\{x: -\pi < x \leqq \pi\}$. It might help to visualize a tape measure that is exactly 2π units long, and thus is just exactly long enough to wrap completely around the circle without any overlap. Remember also that according to this tape measure the radius of the circle is exactly one unit (Figure 8-25).

We are now ready to define the radian measure of $\angle AVB$. When the line segment $\{x: -\pi < x \leqq \pi\}$ is wrapped around the unit

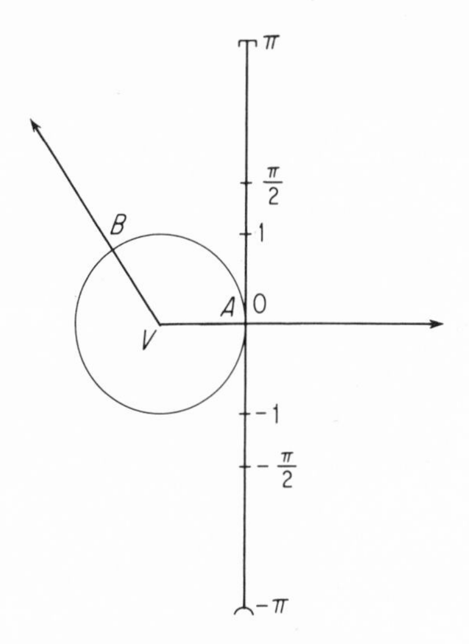

Figure 8-25

circle (with the zero point on the segment held fixed at point A on the circle), there is a unique point on the segment that is mapped onto point B. Suppose that t is the coordinate of this point on the segment that is mapped onto B. Then the radian measure of $\angle AVB$ is $|t|$. This same number, $|t|$, is called the *length of the arc AB intercepted by* $\angle AVB$ *on the unit circle. Thus the radian measure of an angle is the same as the length of the arc that the angle intercepts on a unit circle that has its center at the vertex of the angle.*

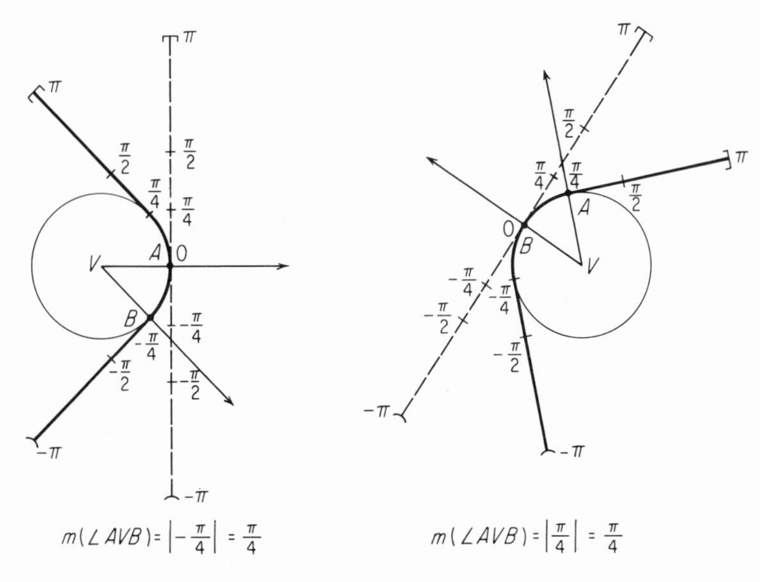

Figure 8-26 (a)

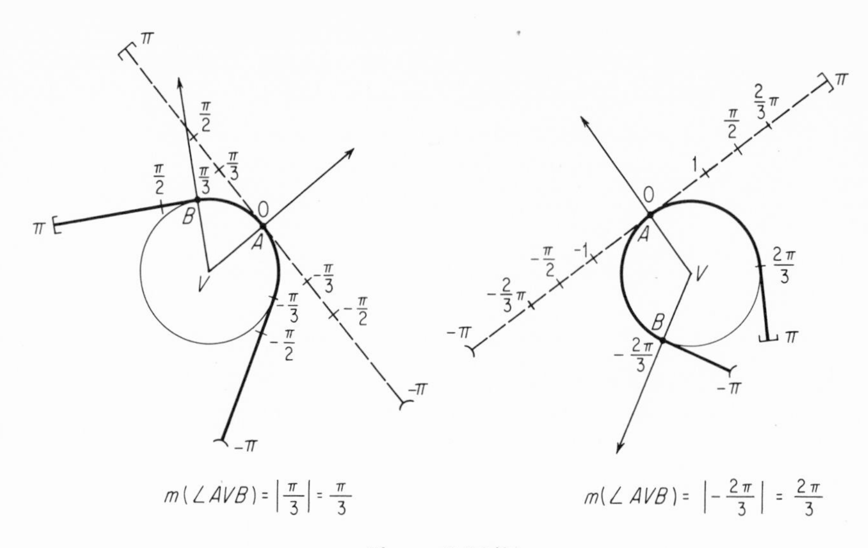

Figure 8-26 (b)

Figure 8-26 shows how the radian measure of four angles may be determined. The symbol $m(\angle AVB)$ is used to denote the radian measure of $\angle AVB$.

Since t is a number such that

$$-\pi < t \leqq \pi$$

it follows that

$$0 \leqq |t| \leqq \pi$$

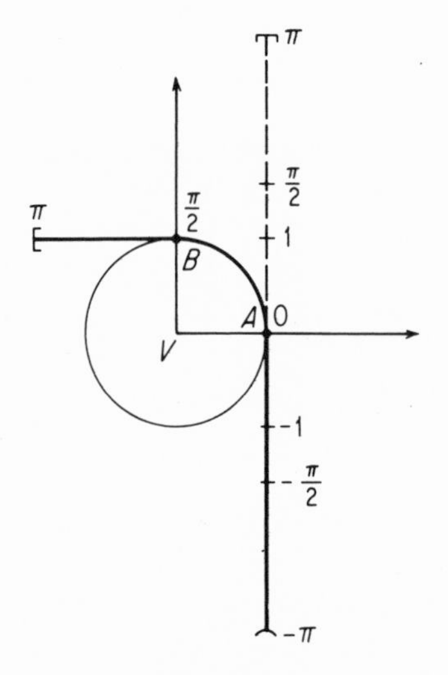

Figure 8-27

This means that the radian measure of a particular angle is not less than 0 and is not greater than π.

If the radian measure of an angle is $\pi/2$, then the length of the arc intercepted by the angle on a unit circle is $\pi/2$, which is one-fourth the circumference of the circle. The angle is a right angle (Figure 8-27).

If the radian measure of an angle is π, then the length of the arc intercepted by the angle on a unit circle is π, which is one-half the circumference of the unit circle. Such an angle is formed by two rays that have a common endpoint and are in the same straight line but have opposite directions (Figure 8-28).

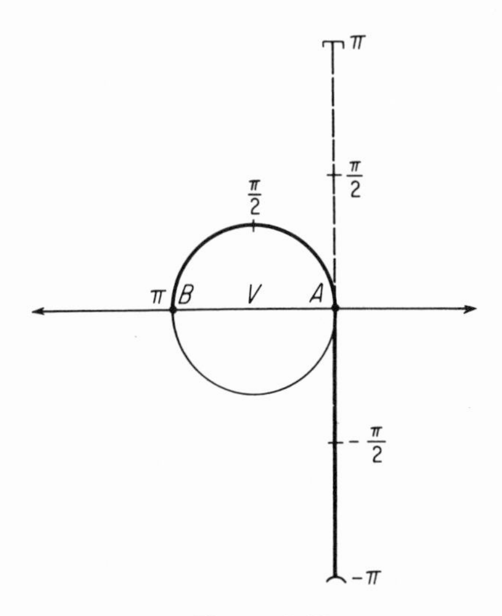

Figure 8-28

Probably the reader is familiar with the degree measure of angles. In Figure 8-29 suppose that the circle is a unit circle with center at the vertex of the angle. Let the length of the arc intercepted by the angle be denoted by s. If s is $\frac{1}{360}$ of the circumference of the circle,

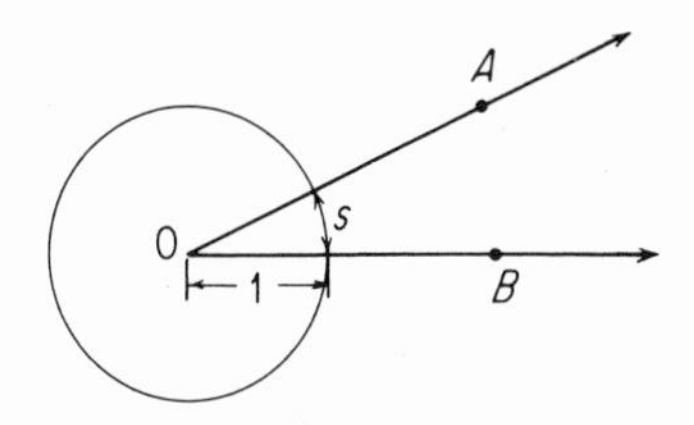

Figure 8-29

then the angle is said to measure 1°. Since the circumference of a unit circle is 2π, this means that if $s = (\frac{1}{360})2\pi = \pi/180$, then a measure of the angle is 1°. In general, if s is the length of the arc of the unit circle intercepted by the angle, the degree measure of the angle is defined to be $(180/\pi)s$.

This definition establishes a rule for a linear function that tells how to convert from radian measure to degree measure:

$$d(r) = \frac{180}{\pi} r$$

where $d(r)$ denotes the degree measure of an angle that measures r radians. Thus:

$$d(1) = \frac{180}{\pi} \approx 57.29578$$

If an angle measures 1 radian, then it measures approximately 57.29578°. Also

$$d\left(\frac{\pi}{2}\right) = \frac{180}{\pi} \cdot \frac{\pi}{2} = 90.$$

If an angle measures $\pi/2$ radians, then it measures 90°.

$$d\left(\frac{\pi}{6}\right) = 30.$$

An angle that measures $\pi/6$ radians also measures 30°.

8-10 EXERCISES

1. Convert from radian measure to degree measure for an angle that measures:

(a) 2 radians
(b) 3 radians
(c) 1.5 radians
(d) $\pi/6$ radians
(e) $\pi/3$ radians
(f) $3\pi/2$ radians

2. Write the rule for the function that converts degree measure of an angle to radian measure: $r(d) = ?$

3. Convert from degree measure to radian measure for an angle that measures:

(a) 45°
(b) 135°
(c) 2°
(d) 120°
(e) 150°
(f) 10°

4. (a) In a circle of radius 10 inches a central angle AOB measures 2 radians. What is the length of the arc of the circle that is intercepted by the angle?
(b) Justify the formula $s = r\theta$, where r denotes the radius of a circle, θ is the radian measure of a central angle, and s is the length of the arc intercepted by the angle.

†**5.** Consider the following method of "measuring" an arbitrary angle AOB. Choose a unit length and then choose point P on the ray OA and point Q on the ray OB such that the line segments OP and OQ are each one unit long.

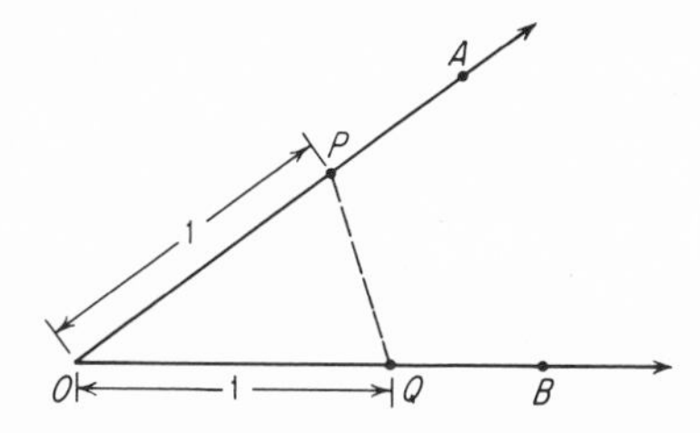

Define the "line segment measure" of angle AOB to be the length of line segment PQ.
(a) Give an advantage of measuring angles in this way.
(b) Give a serious disadvantage of measuring angles in this way.

8-11 THE TRIGONOMETRY OF RIGHT TRIANGLES

Suppose a vertical tree is standing on level ground (Figure 8-30). Point A is 100 feet from the base of the tree. The top of the tree is at

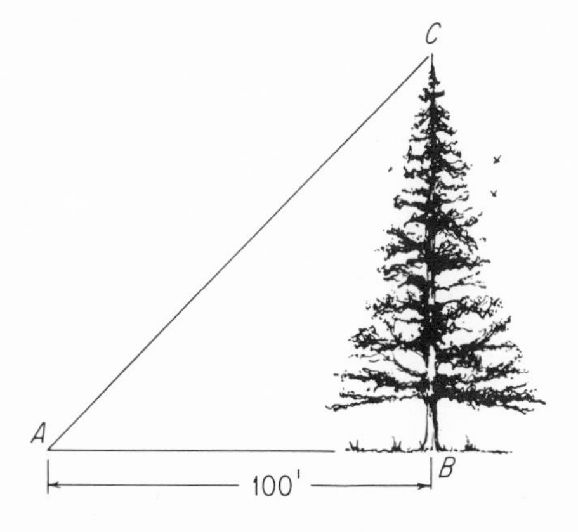

Figure 8-30

C. Angle BAC is measured and found to be 0.8 radians. From this information can we calculate the height of the tree?

The answer is easy to give. Let h denote the height of the tree in feet. Then (for a reason that we shall justify presently)

$$\frac{h}{100} = \tan 0.8$$

Therefore,

$$h = 100 \tan 0.8$$
$$\approx 103$$

The tree is approximately 103 feet high.

This problem is typical of a large class of problems that arise in surveying, navigation, and physics: Certain facts are known about a triangle and from these other facts about the triangle must be deduced. A large part of traditional trigonometry is concerned with the solution of triangles. We shall limit our discussion to trigonometry of right triangles.

When the sine and cosine functions were defined in Section 8-3, no explicit reference was made to right triangles. Sin r and cos r were defined by reference to a unit circle in a coordinate plane. However, if we restrict the domain of these functions to the interval $0 < r < \pi/2$, then definitions that are equivalent to those of Section 8-3 can be made by a somewhat different approach, using the lengths of the sides of a right triangle. We begin with a statement of the principal facts to be developed.

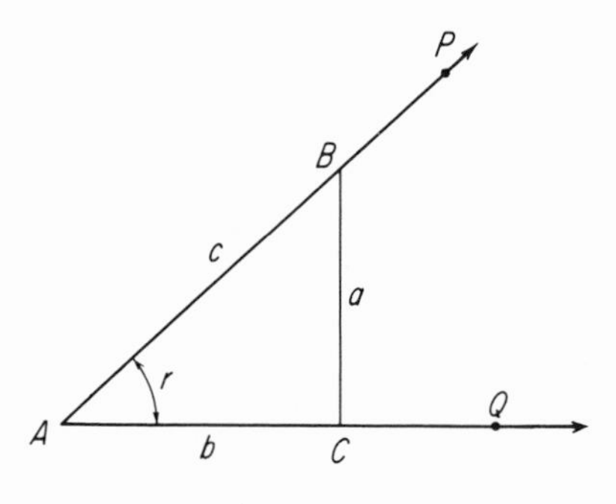

Figure 8-31

Let r denote any number between 0 and $\pi/2$. Then there is an angle PAQ that has radian measure r (Figure 8-31). Select an arbitrary point B (not the same as A) on one of the rays that form this angle; let C denote the point on the ray AQ such that $\angle ACB$ is a right angle. In the right triangle ABC let a denote the length of the side opposite $\angle A$, let b denote the length of the side opposite $\angle B$, and let c denote the length of the side opposite $\angle C$. Then:

$$\sin r = \frac{a}{c} \tag{1}$$

$$\cos r = \frac{b}{c} \tag{2}$$

$$\tan r = \frac{a}{b} \tag{3}$$

(The third one of these facts was used in solving the "tree" exercise at the beginning of this section.)

In order to justify these facts, construct a rectangular coordinate system that has its origin at A, the positive x-axis lying along one of the rays of the angle, and the positive y-axis chosen so that the other ray lies in the first quadrant. (This can be done since we are restricting our discussion to those angles A such that the radian measure of A is between 0 and $\pi/2$.) Unit length on each axis is chosen to be the same as that used in measuring the sides of the triangle (Figure 8-32).

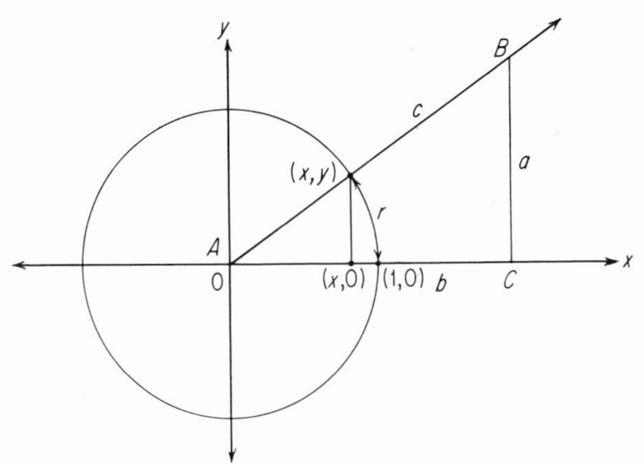

Figure 8-32

The unit circle with center at the origin intersects the ray AB in a point whose coordinates we shall call (x, y). Then the radian measure of angle A is the length of the arc of the circle that extends from $(1, 0)$ to (x, y). Denote the length of this arc by r. Then $\sin r = y$ and $\cos r = x$ by our basic definitions. At this point we appeal to a fundamental theorem from high school geometry: Triangle ABC is similar to the right triangle whose vertices are A, the point with coordinates (x, y), and the point with coordinates $(x, 0)$. The lengths of corresponding sides of similar triangles are proportional. Therefore,

$$\frac{y}{1} = \frac{a}{c} \quad \text{and} \quad \frac{x}{1} = \frac{b}{c}.$$

This means

$$\sin r = y = \frac{a}{c} \quad \text{and} \quad \cos r = x = \frac{b}{c}.$$

Finally,

$$\tan r = \frac{\sin r}{\cos r} = \frac{a}{b}.$$

The three facts established here are worth repeating in a slightly different form and from a different point of view. Every right triangle ABC (with side BC perpendicular to side AC) has three angles:

$$\angle A = \angle BAC$$

$$\angle B = \angle ABC$$

$$\angle C = \angle ACB$$

The radian measure of $\angle C$ is $\pi/2$ (Figure 8-33).

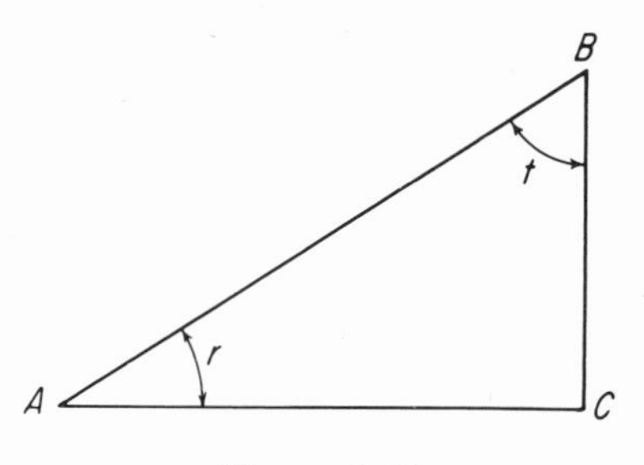

Figure 8-33

By definition, the radian measure of an angle is a number that is not negative. A basic theorem from high school geometry tells us that the sum of the radian measures of these three angles is π. The measure of angle C is $\pi/2$; therefore, the measure of angle A (like that of angle B) is between 0 and $\pi/2$. Such angles are called **acute angles.**

Now let r denote the radian measure of angle A. Then:

$$\frac{\text{length of side opposite } A}{\text{length of hypotenuse}} = \sin r$$

$$\frac{\text{length of side adjacent to } A}{\text{length of hypotenuse}} = \cos r$$

$$\frac{\text{length of side opposite } A}{\text{length of side adjacent to } A} = \tan r$$

If we let t denote the radian measure of angle B, then:

$$\frac{\text{length of side opposite } B}{\text{length of hypotenuse}} = \sin t$$

$$\frac{\text{length of side adjacent to } B}{\text{length of hypotenuse}} = \cos t$$

$$\frac{\text{length of side opposite } B}{\text{length of side adjacent to } B} = \tan t$$

The reader should observe that $r + t = \pi/2$, since we have assumed that the sum of the radian measures of the three angles is π.

Angles are frequently measured in degrees rather than in radians. It would be convenient if we could use such expressions as $\sin 30°$, $\cos 75°$, and $\tan 16.1°$. This can be done provided we have a clear understanding of what these expressions mean. An angle that measures $x°$ also measures $\pi x/180$ radians. Therefore, we make the following definitions for each number x that lies between 0 and 180:

$$\cos x° = \cos \frac{\pi x}{180}$$

$$\sin x° = \sin \frac{\pi x}{180}$$

$$\tan x° = \tan \frac{\pi x}{180}$$

For example,

$$\sin 30° = \sin \frac{\pi}{6}$$

$$\cos 75° = \cos \frac{5\pi}{12}$$

and

$$\tan 16.1° = \tan \frac{16.1\pi}{180}.$$

Table 3 in the Appendix enables us to approximate sin 27°, cos 78°, tan 16.1°, etc., without first converting from the degree measure of an angle to the radian measure. Also, this table can be used to approximate the degree measure of an acute angle of a right triangle if we know the lengths of two sides of the triangle.

EXAMPLE 1: Suppose the sides of right triangle ABC are 3, 4, and 5 units in length, as designated in Figure 8-34. Angle C

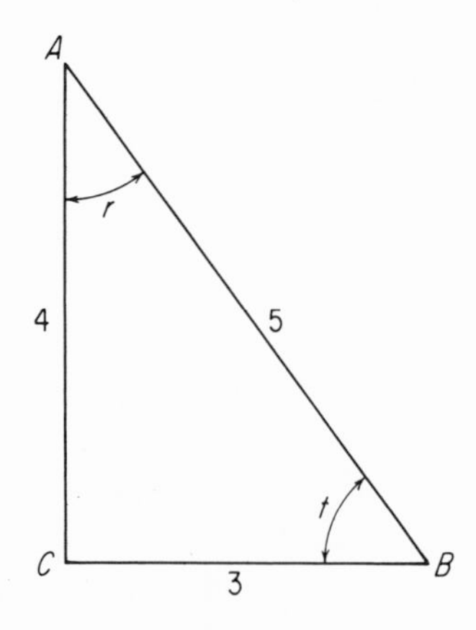

Figure 8-34

measures $\pi/2$ radians. Let r and t denote the radian measures of angles A and B. What are $\sin r$, $\cos r$, $\tan r$, $\sin t$, $\cos t$, and $\tan t$? Use Table 2 to approximate r and t (accurate to three significant digits). Then use Table 3 to approximate the degree measures of angles A and B (accurate to the nearest degree).

Solution:

$$\sin r = \tfrac{3}{5} \qquad \sin t = \tfrac{4}{5}$$

$$\cos r = \tfrac{4}{5} \qquad \cos t = \tfrac{3}{5}$$

$$\tan r = \tfrac{3}{4} \qquad \tan t = \tfrac{4}{3}$$

Since $\sin r = \frac{3}{5} = 0.6$, look in Table 2 to locate a number whose sine is 0.6. We find that $r \approx 0.644$. To find t, observe that $r + t = \pi/2$ because the sum of the measures of the three angles is π and angle C measures $\pi/2$ radians. Therefore,

$$t = (\pi/2) - r \approx 0.927.$$

From Table 3 we find that angle A measures approximately 37°. Therefore, angle B measures approximately 53°.

EXAMPLE 2: Suppose it is known that in triangle ABC, angle A measures 25°, angle C is a right angle, and side BC is 10 units long (Figure 8-35). Solve the triangle; that is, find the lengths of

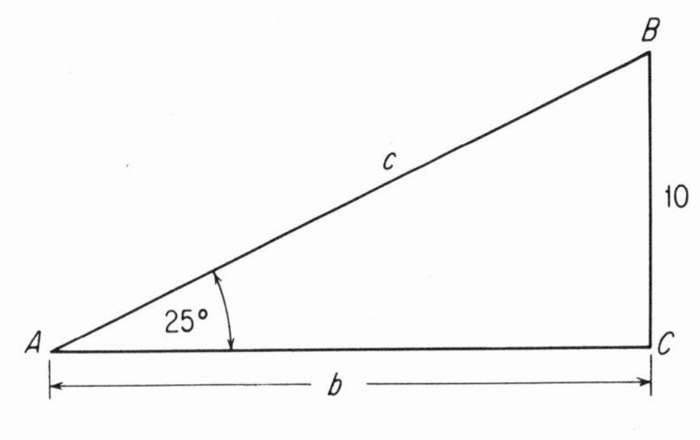

Figure 8-35

all the sides and the measures of all the angles. Express the lengths of the sides accurate to two significant digits.

Solution:

(1) The measure of angle B is 65°, since the sum of the measures of angles A and B is 90°.

(2) $b/10 = \tan 65°$. Therefore, $b = 10 \tan 65° \approx 21$.

(3) $10/c = \sin 25°$. Therefore,

$$c = 10/(\sin 25°) \approx 10/0.4226 \approx 24.$$

EXAMPLE 3: A flagpole that is 40 feet high casts a shadow 25 feet long. What is the measure (in radians and also in degrees) of

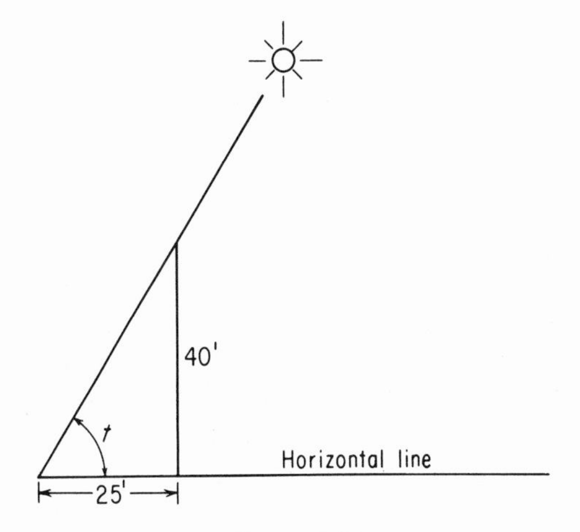

Figure 8-36

the angle of elevation of the sun? Give the radian measure accurate to the nearest tenth of a radian and the degree measure accurate to the nearest degree.

Solution: Let t denote the radian measure of the angle of elevation of the sun. Then $\tan t = \frac{40}{25} = 1.6$. From Table 2 we find that the angle of elevation measures about 1.0 radians. This is approximately 57°.

EXAMPLE 4: A ship is sailing due east, parallel to the shore line, at 12 miles per hour. At 6:00 A.M. a light on shore is

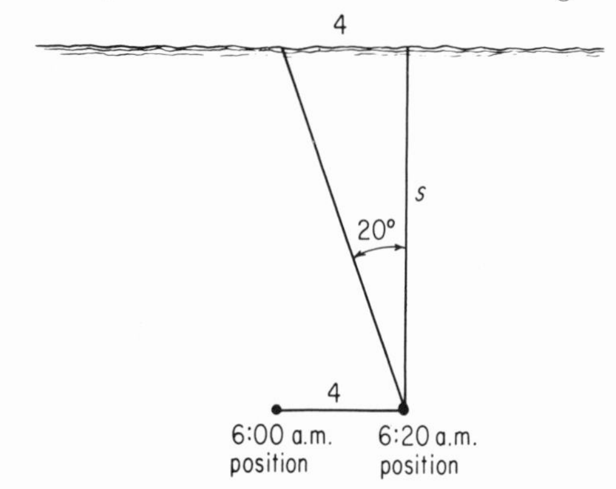

Figure 8-37

sighted due north. At 6:20 A.M. the light is bearing twenty degrees west of North. How far offshore is the ship? Give the answer accurate to the nearest mile.

Solution: We need to find the length of a side of a right triangle, given the length of another side and the measure of one of the acute angles. We know that $s/4 = \tan 70°$. Therefore, $s = 4 \tan 70° \approx 10.99$. The ship is approximately 11 miles offshore.

8-11 EXERCISES

1. The sides of a right triangle are 5, 12, and 13 units long.
(a) Use Table 2 to approximate the radian measures of the two acute angles associated with the triangle. Express the answers accurate to two significant digits.

(b) Use Table 3 to approximate (to the nearest degree) the degree measure of the same two angles.

2. The hypotenuse of a right triangle is 20 units long and one of the other sides is 12 units long. Use Table 3 to approximate (to the nearest degree) the degree measure of the angle determined by these two sides.

3. Consider the right triangle whose vertices are located at (0, 0), (3, 5), and (3, 0) in a rectangular coordinate plane. Use Table 3 to approximate (to the nearest degree) the degree measure of each of the acute angles associated with this triangle.

4. In a rectangular coordinate plane a right triangle has vertices at (0, 0), (0, 4), and (2, 4).

(a) How long is the hypotenuse? Express the answer accurate to the nearest tenth of a unit.

(b) Use Table 3 to approximate (to the nearest degree) the degree measure of the two acute angles associated with this triangle.

5. At a point 300.0 ft from the base of a silo the measure of the angle between the horizontal (line of sight to the base) and the line to the top of the silo is 30.0°. What is the height of the silo? Express the answer accurate to the nearest foot.

6. To measure indirectly the width of a river, a stake was driven into the ground on the north bank directly north of a tree on the opposite bank. From a point 2.00×10^2 ft due east of the stake the tree was sighted and the angle

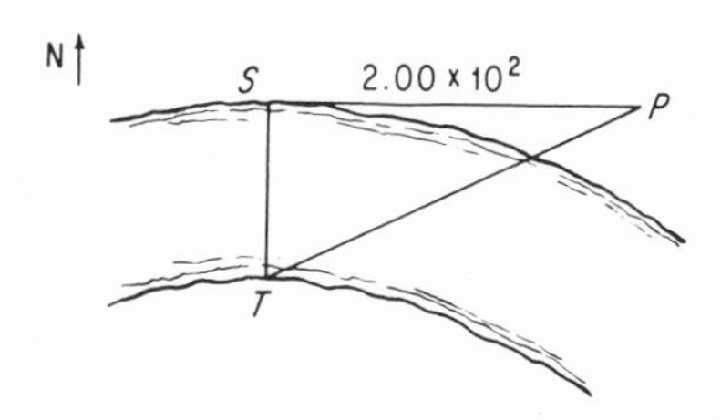

between the line of sight and the east-west line was measured. What is the width of the river if the measure of this angle was 30.0°? Express the answer to the nearest foot.

7. A wire 50.0 ft long is stretched from ground level to the top of a 25-foot pole.

(a) Find the degree measure (accurate to the nearest degree) of the angle between the pole and the wire.

(b) Find the degree measure (accurate to the nearest degree) of the angle between the ground and the wire.

8. A roadbed rises 15.0 ft in 1.000×10^3 ft along the roadbed. What is the approximate degree measure of the angle of inclination of the road to the horizontal? (Express your answer accurate to the nearest degree.)

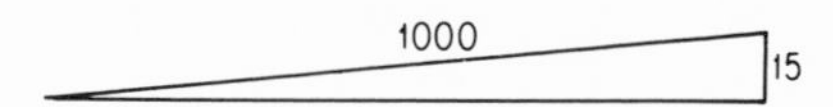

9. From the top of a cliff 150 ft high an observer finds that the "angle of depression" (angle *ABC*) of a steamship at sea measures 15°. How far out from the base of the cliff is the steamship? Express your answer accurate to the nearest hundred feet.

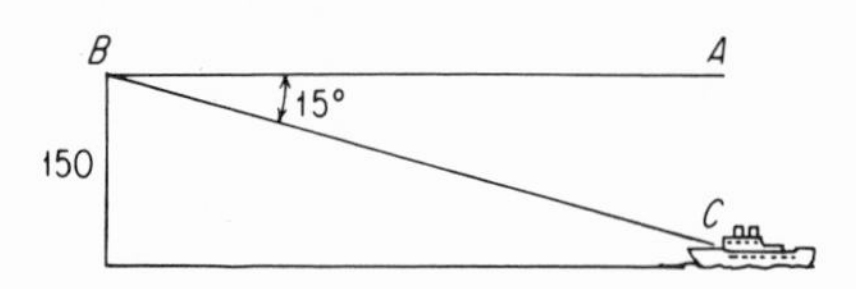

10. A ship is heading due north at a speed that would be 20 ± 0.5 miles per hour if there were no current. However, there is a current of 3.0 miles per hour moving due east.

(a) What is the actual speed of the ship? Express the answer accurate to two significant digits.

(b) Clearly, the ship moves in a direction somewhat east of north. Find this direction (accurate to the nearest degree).

11. A flagpole causes a shadow 40.0 ft long when the elevation of the sun measures 50.0°. How tall is the flagpole? Express the answer accurate to three significant digits.

12. A pilot in a plane with a cruising speed of 200 miles per hour sets his course due west. After flying 30 minutes he discovers that he is 15 miles west and 20 miles north of where he would have been if there had been no wind. Assuming that the wind has been constant during this period, calculate its speed and direction. (Express the calculated wind speed accurate to two significant digits, and the calculated direction to the nearest degree.)

13. Suppose a projectile is to be fired from a gun with an initial velocity v. Let r denote the radian measure of the angle of elevation of the gun; let D

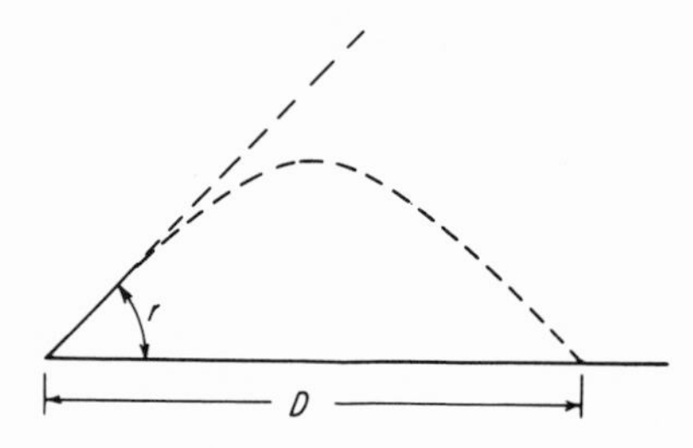

denote the distance from the point where the gun is fired to the point where the projectile returns to the ground. Then D is a function of r, and it can be shown that if air friction is not taken into account, $D(r) = (v^2/32) \sin (2r)$, where v is measured in feet per second and D is measured in feet.

(a) Calculate D if $v = 1.60 \times 10^3$ feet per second and $r = \pi/12.0$. Express the answer to three significant digits.

(b) For a given initial velocity, is there an angle of elevation that will maximize D? If so, what is its measure?

Bibliography

1. Allendoerfer, Carl B., and Cletus O. Oakley, *Fundamentals of College Algebra.* New York: McGraw-Hill Book Co., 1967.
2. Bristol, James D., *The Concept of a Function.* Boston: D. C. Heath and Company, 1963.
3. Drobot, Stefan, *Real Numbers.* Englewood Cliffs, N. J.: Prentice-Hall, Inc., 1964.
4. Fleenor, Charles R., Merrill E. Shanks, and Charles F. Brumfiel, *The Elementary Functions.* Reading, Massachusetts: Addison-Wesley Publishing Company, 1968.
5. Meserve, Bruce E., and Max A. Sobel, *Mathematics for Secondary School Teachers.* Englewood Cliffs, N. J.: Prentice-Hall, Inc., 1962.
6. Niven, Ivan, *Numbers: Rational and Irrational.* New York: Random House, 1961.
7. Parker, Francis D., *The Structure of Number Systems.* Englewood Cliffs, N. J.: Prentice-Hall, Inc., 1966.
8. School Mathematics Study Group, *Elementary Functions.* New Haven and London: Yale University Press, 1961.
9. Smith, Seaton E., Jr., *Explorations in Elementary Mathematics.* Englewood Cliffs, N. J.: Prentice-Hall, Inc., 1966.
10. Smith, William K., *Inverse Functions.* New York: The Macmillan Company, 1966.
11. Yandl, Andre L., *The Non-Algebraic Elementary Functions.* Englewood Cliffs, N. J.: Prentice-Hall, Inc., 1964.

Appendix

TABLE 1. Common Logarithms of Numbers

N	0	1	2	3	4	5	6	7	8	9
1.0	.0000	.0043	.0086	.0128	.0170	.0212	.0253	.0294	.0334	.0374
1.1	.0414	.0453	.0492	.0531	.0569	.0607	.0645	.0682	.0719	.0755
1.2	.0792	.0828	.0864	.0899	.0934	.0969	.1004	.1038	.1072	.1106
1.3	.1139	.1173	.1206	.1239	.1271	.1303	.1335	.1367	.1399	.1430
1.4	.1461	.1492	.1523	.1553	.1584	.1614	.1644	.1673	.1703	.1732
1.5	.1761	.1790	.1818	.1847	.1875	.1903	.1931	.1959	.1987	.2014
1.6	.2041	.2068	.2095	.2122	.2148	.2175	.2201	.2227	.2253	.2279
1.7	.2304	.2330	.2355	.2380	.2405	.2430	.2455	.2480	.2504	.2529
1.8	.2553	.2577	.2601	.2625	.2648	.2672	.2695	.2718	.2742	.2765
1.9	.2788	.2810	.2833	.2856	.2878	.2900	.2923	.2945	.2967	.2989
2.0	.3010	.3032	.3054	.3075	.3096	.3118	.3139	.3160	.3181	.3201
2.1	.3222	.3243	.3263	.3284	.3304	.3324	.3345	.3365	.3385	.3404
2.2	.3424	.3444	.3464	.3483	.3502	.3522	.3541	.3560	.3579	.3598
2.3	.3617	.3636	.3655	.3674	.3692	.3711	.3729	.3747	.3766	.3784
2.4	.3802	.3820	.3838	.3856	.3874	.3892	.3909	.3927	.3945	.3962
2.5	.3979	.3997	.4014	.4031	.4048	.4065	.4082	.4099	.4116	.4133
2.6	.4150	.4166	.4183	.4200	.4216	.4232	.4249	.4265	.4281	.4298
2.7	.4314	.4330	.4346	.4362	.4378	.4393	.4409	.4425	.4440	.4456
2.8	.4472	.4487	.4502	.4518	.4533	.4548	.4564	.4579	.4594	.4609
2.9	.4624	.4639	.4654	.4669	.4683	.4698	.4713	.4728	.4742	.4757
3.0	.4771	.4786	.4800	.4814	.4829	.4843	.4857	.4871	.4886	.4900
3.1	.4914	.4928	.4942	.4955	.4969	.4983	.4997	.5011	.5024	.5038
3.2	.5051	.5065	.5079	.5092	.5105	.5119	.5132	.5145	.5159	.5172
3.3	.5185	.5198	.5211	.5224	.5237	.5250	.5263	.5276	.5289	.5302
3.4	.5315	.5328	.5340	.5353	.5366	.5378	.5391	.5403	.5416	.5428
3.5	.5441	.5453	.5465	.5478	.5490	.5502	.5514	.5527	.5539	.5551
3.6	.5563	.5575	.5587	.5599	.5611	.5623	.5635	.5647	.5658	.5670
3.7	.5682	.5694	.5705	.5717	.5729	.5740	.5752	.5763	.5775	.5786
3.8	.5798	.5809	.5821	.5832	.5843	.5855	.5866	.5877	.5888	.5899
3.9	.5911	.5922	.5933	.5944	.5955	.5966	.5977	.5988	.5999	.6010
4.0	.6021	.6031	.6042	.6053	.6064	.6075	.6085	.6096	.6107	.6117
4.1	.6128	.6138	.6149	.6160	.6170	.6180	.6191	.6201	.6212	.6222
4.2	.6232	.6243	.6253	.6263	.6274	.6284	.6294	.6304	.6314	.6325
4.3	.6335	.6345	.6355	.6365	.6375	.6385	.6395	.6405	.6415	.6425
4.4	.6435	.6444	.6454	.6464	.6474	.6484	.6493	.6503	.6513	.6522
4.5	.6532	.6542	.6551	.6561	.6571	.6580	.6590	.6599	.6609	.6618
4.6	.6628	.6637	.6646	.6656	.6665	.6675	.6684	.6693	.6702	.6712
4.7	.6721	.6730	.6739	.6749	.6758	.6767	.6776	.6785	.6794	.6803
4.8	.6812	.6821	.6830	.6839	.6848	.6857	.6866	.6875	.6884	.6893
4.9	.6902	.6911	.6920	.6928	.6937	.6946	.6955	.6964	.6972	.6981
5.0	.6990	.6998	.7007	.7016	.7024	.7033	.7042	.7050	.7059	.7067
5.1	.7076	.7084	.7093	.7101	.7110	.7118	.7126	.7135	.7143	.7152
5.2	.7160	.7168	.7177	.7185	.7193	.7202	.7210	.7218	.7226	.7235
5.3	.7243	.7251	.7259	.7267	.7275	.7284	.7292	.7300	.7308	.7316
5.4	.7324	.7332	.7340	.7348	.7356	.7364	.7372	.7380	.7388	.7396
N	0	1	2	3	4	5	6	7	8	9

TABLE 1. (CONTINUED)

N	0	1	2	3	4	5	6	7	8	9
5.5	.7404	.7412	.7419	.7427	.7435	.7443	.7451	.7459	.7466	.7474
5.6	.7482	.7490	.7497	.7505	.7513	.7520	.7528	.7536	.7543	.7551
5.7	.7559	.7566	.7574	.7582	.7589	.7597	.7604	.7612	.7619	.7627
5.8	.7634	.7642	.7649	.7657	.7664	.7672	.7679	.7686	.7694	.7701
5.9	.7709	.7716	.7723	.7731	.7738	.7745	.7752	.7760	.7767	.7774
6.0	.7782	.7789	.7796	.7803	.7810	.7818	.7825	.7832	.7839	.7846
6.1	.7853	.7860	.7868	.7875	.7882	.7889	.7896	.7903	.7910	.7917
6.2	.7924	.7931	.7938	.7945	.7952	.7959	.7966	.7973	.7980	.7987
6.3	.7993	.8000	.8007	.8014	.8021	.8028	.8035	.8041	.8048	.8055
6.4	.8062	.8069	.8075	.8082	.8089	.8096	.8102	.8109	.8116	.8122
6.5	.8129	.8136	.8142	.8149	.8156	.8162	.8169	.8176	.8182	.8189
6.6	.8195	.8202	.8209	.8215	.8222	.8228	.8235	.8241	.8248	.8254
6.7	.8261	.8267	.8274	.8280	.8287	.8293	.8299	.8306	.8312	.8319
6.8	.8325	.8331	.8338	.8344	.8351	.8357	.8363	.8370	.8376	.8382
6.9	.8388	.8395	.8401	.8407	.8414	.8420	.8426	.8432	.8439	.8445
7.0	.8451	.8457	.8463	.8470	.8476	.8482	.8488	.8494	.8500	.8506
7.1	.8513	.8519	.8525	.8531	.8537	.8543	.8549	.8555	.8561	.8567
7.2	.8573	.8579	.8585	.8591	.8597	.8603	.8609	.8615	.8621	.8627
7.3	.8633	.8639	.8645	.8651	.8657	.8663	.8669	.8675	.8681	.8686
7.4	.8692	.8698	.8704	.8710	.8716	.8722	.8727	.8733	.8739	.8745
7.5	.8751	.8756	.8762	.8768	.8774	.8779	.8785	.8791	.8797	.8802
7.6	.8808	.8814	.8820	.8825	.8831	.8837	.8842	.8848	.8854	.8859
7.7	.8865	.8871	.8876	.8882	.8887	.8893	.8899	.8904	.8910	.8915
7.8	.8921	.8927	.8932	.8938	.8943	.8949	.8954	.8960	.8965	.8971
7.9	.8976	.8982	.8987	.8993	.8998	.9004	.9009	.9015	.9020	.9025
8.0	.9031	.9036	.9042	.9047	.9053	.9058	.9063	.9069	.9074	.9079
8.1	.9085	.9090	.9096	.9101	.9106	.9112	.9117	.9122	.9128	.9133
8.2	.9138	.9143	.9149	.9154	.9159	.9165	.9170	.9175	.9180	.9186
8.3	.9191	.9196	.9201	.9206	.9212	.9217	.9222	.9227	.9232	.9238
8.4	.9243	.9248	.9253	.9258	.9263	.9269	.9274	.9279	.9284	.9289
8.5	.9294	.9299	.9304	.9309	.9315	.9320	.9325	.9330	.9335	.9340
8.6	.9345	.9350	.9355	.9360	.9365	.9370	.9375	.9380	.9385	.9390
8.7	.9395	.9400	.9405	.9410	.9415	.9420	.9425	.9430	.9435	.9440
8.8	.9445	.9450	.9455	.9460	.9465	.9469	.9474	.9479	.9484	.9489
8.9	.9494	.9499	.9504	.9509	.9513	.9518	.9523	.9528	.9533	.9538
9.0	.9542	.9547	.9552	.9557	.9562	.9566	.9571	.9576	.9581	.9586
9.1	.9590	.9595	.9600	.9605	.9609	.9614	.9619	.9624	.9628	.9633
9.2	.9638	.9643	.9647	.9652	.9657	.9661	.9666	.9671	.9675	.9680
9.3	.9685	.9689	.9694	.9699	.9703	.9708	.9713	.9717	.9722	.9727
9.4	.9731	.9736	.9741	.9745	.9750	.9754	.9759	.9763	.9768	.9773
9.5	.9777	.9782	.9786	.9791	.9795	.9800	.9805	.9809	.9814	.9818
9.6	.9823	.9827	.9832	.9836	.9841	.9845	.9850	.9854	.9859	.9863
9.7	.9868	.9872	.9877	.9881	.9886	.9890	.9894	.9899	.9903	.9908
9.8	.9912	.9917	.9921	.9926	.9930	.9934	.9939	.9943	.9948	.9952
9.9	.9956	.9961	.9965	.9969	.9974	.9978	.9983	.9987	.9991	.9996
N	0	1	2	3	4	5	6	7	8	9

TABLE 2

x	Sin x	Cos x	Tan x	x	Sin x	Cos x	Tan x
.00	.0000	1.0000	.0000	.41	.3986	.9171	.4346
.01	.0100	1.0000	.0100	.42	.4078	.9131	.4466
.02	.0200	.9998	.0200	.43	.4169	.9090	.4586
.03	.0300	.9996	.0300	.44	.4259	.9048	.4708
.04	.0400	.9992	.0400	.45	.4350	.9004	.4831
.05	.0500	.9988	.0500	.46	.4439	.8961	.4954
.06	.0600	.9982	.0601	.47	.4529	.8916	.5080
.07	.0699	.9976	.0701	.48	.4618	.8870	.5206
.08	.0799	.9968	.0802	.49	.4706	.8823	.5334
.09	.0899	.9960	.0902	.50	.4794	.8776	.5463
.10	.0998	.9950	.1003	.51	.4882	.8727	.5594
.11	.1098	.9940	.1104	.52	.4969	.8678	.5726
.12	.1197	.9928	.1206	.53	.5055	.8628	.5859
.13	.1296	.9916	.1307	.54	.5141	.8577	.5994
.14	.1395	.9902	.1409	.55	.5227	.8525	.6131
.15	.1494	.9888	.1511	.56	.5312	.8473	.6269
.16	.1593	.9872	.1614	.57	.5396	.8419	.6410
.17	.1692	.9856	.1717	.58	.5480	.8365	.6552
.18	.1790	.9838	.1820	.59	.5564	.8309	.6696
.19	.1889	.9820	.1923	.60	.5646	.8253	.6841
.20	.1987	.9801	.2027	.61	.5729	.8196	.6989
.21	.2085	.9780	.2131	.62	.5810	.8139	.7139
.22	.2182	.9759	.2236	.63	.5891	.8080	.7291
.23	.2280	.9737	.2341	.64	.5972	.8021	.7445
.24	.2377	.9713	.2447	.65	.6052	.7961	.7602
.25	.2474	.9689	.2553	.66	.6131	.7900	.7761
.26	.2571	.9664	.2660	.67	.6210	.7838	.7923
.27	.2667	.9638	.2768	.68	.6288	.7776	.8087
.28	.2764	.9611	.2876	.69	.6365	.7712	.8253
.29	.2860	.9582	.2984	.70	.6442	.7648	.8423
.30	.2955	.9553	.3093	.71	.6518	.7584	.8595
.31	.3051	.9523	.3203	.72	.6594	.7518	.8771
.32	.3146	.9492	.3314	.73	.6669	.7452	.8949
.33	.3240	.9460	.3425	.74	.6743	.7385	.9131
.34	.3335	.9428	.3537	.75	.6816	.7317	.9316
.35	.3429	.9394	.3650	.76	.6889	.7248	.9505
.36	.3523	.9359	.3764	.77	.6961	.7179	.9697
.37	.3616	.9323	.3879	.78	.7033	.7109	.9893
.38	.3709	.9287	.3994	.79	.7104	.7038	1.009
.39	.3802	.9249	.4111	.80	.7174	.6967	1.030
.40	.3894	.9211	.4228				
x	Sin x	Cos x	Tan x	x	Sin x	Cos x	Tan x

TABLE 2. (CONTINUED)

x	Sin x	Cos x	Tan x	x	Sin x	Cos x	Tan x
.81	.7243	.6895	1.050	1.21	.9356	.3530	2.650
.82	.7311	.6822	1.072	1.22	.9391	.3436	2.733
.83	.7379	.6749	1.093	1.23	.9425	.3342	2.820
.84	.7446	.6675	1.116	1.24	.9458	.3248	2.912
.85	.7513	.6600	1.138	1.25	.9490	.3153	3.010
.86	.7578	.6524	1.162	1.26	.9521	.3058	3.113
.87	.7643	.6448	1.185	1.27	.9551	.2963	3.224
.88	.7707	.6372	1.210	1.28	.9580	.2867	3.341
.89	.7771	.6294	1.235	1.29	.9608	.2771	3.467
.90	.7833	.6216	1.260	1.30	.9636	.2675	3.602
.91	.7895	.6137	1.286	1.31	.9662	.2579	3.747
.92	.7956	.6058	1.313	1.32	.9687	.2482	3.903
.93	.8016	.5978	1.341	1.33	.9711	.2385	4.072
.94	.8076	.5898	1.369	1.34	.9735	.2288	4.256
.95	.8134	.5817	1.398	1.35	.9757	.2190	4.455
.96	.8192	.5735	1.428	1.36	.9779	.2092	4.673
.97	.8249	.5653	1.459	1.37	.9799	.1994	4.913
.98	.8305	.5570	1.491	1.38	.9819	.1896	5.177
.99	.8360	.5487	1.524	1.39	.9837	.1798	5.471
1.00	.8415	.5403	1.557	1.40	.9854	.1700	5.798
1.01	.8468	.5319	1.592	1.41	.9871	.1601	6.165
1.02	.8521	.5234	1.628	1.42	.9887	.1502	6.581
1.03	.8573	.5148	1.665	1.43	.9901	.1403	7.055
1.04	.8624	.5062	1.704	1.44	.9915	.1304	7.602
1.05	.8674	.4976	1.743	1.45	.9927	.1205	8.238
1.06	.8724	.4889	1.784	1.46	.9939	.1106	8.989
1.07	.8772	.4801	1.827	1.47	.9949	.1006	9.887
1.08	.8820	.4713	1.871	1.48	.9959	.0907	10.98
1.09	.8866	.4625	1.917	1.49	.9967	.0807	12.35
1.10	.8912	.4536	1.965	1.50	.9975	.0707	14.10
1.11	.8957	.4447	2.014	1.51	.9982	.0608	16.43
1.12	.9001	.4357	2.066	1.52	.9987	.0508	19.67
1.13	.9044	.4267	2.120	1.53	.9992	.0408	24.50
1.14	.9086	.4176	2.176	1.54	.9995	.0308	32.46
1.15	.9128	.4085	2.234	1.55	.9998	.0208	48.08
1.16	.9168	.3993	2.296	1.56	.9999	.0108	92.62
1.17	.9208	.3902	2.360	1.57	1.000	.0008	1256
1.18	.9246	.3809	2.427	1.58	1.000	−.0092	−108.7
1.19	.9284	.3717	2.498	1.59	.9998	−.0192	−52.07
1.20	.9320	.3624	2.572	1.60	.9996	−.0292	−34.23

TABLE 3

$x°$	Sin $x°$	Cos $x°$	Tan $x°$	Cot $x°$	
0	.0000	1.0000	.0000	. . .	90
1	.0175	.9998	.0175	57.290	89
2	.0349	.9994	.0349	28.636	88
3	.0523	.9986	.0524	19.081	87
4	.0698	.9976	.0699	14.301	86
5	.0872	.9962	.0875	11.430	85
6	.1045	.9945	.1051	9.514	84
7	.1219	.9925	.1228	8.144	83
8	.1392	.9903	.1405	7.115	82
9	.1564	.9877	.1584	6.314	81
10	.1736	.9848	.1763	5.671	80
11	.1908	.9816	.1944	5.145	79
12	.2079	.9781	.2126	4.705	78
13	.2250	.9744	.2309	4.332	77
14	.2419	.9703	.2493	4.011	76
15	.2588	.9659	.2679	3.732	75
16	.2756	.9613	.2867	3.487	74
17	.2924	.9563	.3057	3.271	73
18	.3090	.9511	.3249	3.078	72
19	.3256	.9455	.3443	2.904	71
20	.3420	.9397	.3640	2.748	70
21	.3584	.9336	.3839	2.605	69
22	.3746	.9272	.4040	2.475	68
23	.3907	.9205	.4245	2.356	67
24	.4067	.9135	.4452	2.246	66
25	.4226	.9063	.4663	2.145	65
26	.4384	.8988	.4877	2.050	64
27	.4540	.8910	.5095	1.963	63
28	.4695	.8829	.5317	1.881	62
29	.4848	.8746	.5543	1.804	61
30	.5000	.8660	.5774	1.732	60
31	.5150	.8572	.6009	1.664	59
32	.5299	.8480	.6249	1.600	58
33	.5446	.8387	.6494	1.540	57
34	.5592	.8290	.6745	1.483	56
35	.5736	.8192	.7002	1.428	55
36	.5878	.8090	.7265	1.376	54
37	.6018	.7986	.7536	1.327	53
38	.6157	.7880	.7813	1.280	52
39	.6293	.7771	.8098	1.235	51
40	.6428	.7660	.8391	1.192	50
41	.6561	.7547	.8693	1.150	49
42	.6691	.7431	.9004	1.111	48
43	.6820	.7314	.9325	1.072	47
44	.6947	.7193	.9657	1.036	46
45	.7071	.7071	1.0000	1.000	45
	Cos $x°$	Sin $x°$	Cot $x°$	Tan $x°$	$x°$

Answers to Odd-Numbered Exercises

CHAPTER 1—NUMBER LINES AND COORDINATE PLANES

1-2 Some Subsets of a Real Number Line

1. 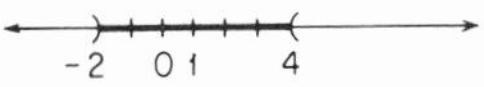 -2 0 1 4

3. 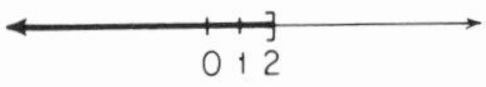 0 1 2

5. 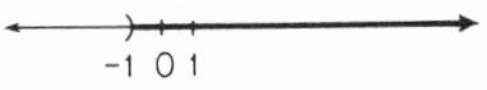 -1 0 1

7. 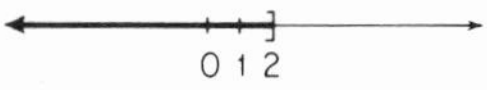 0 1 2

9. 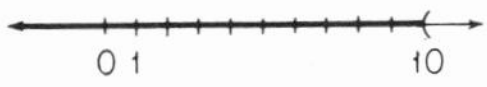 0 1 10

11. 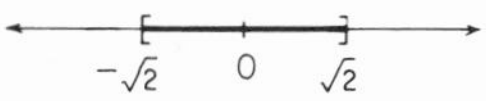 $-\sqrt{2}$ 0 $\sqrt{2}$

1-3 The Absolute Value of a Number

1. (a) $8, -8$; (b) $-6, 14$; (c) $-10, 2$; (d) $-10, -4$; (e) $\frac{7}{2}, -\frac{5}{2}$; (f) $-2, 4$; (g) $-4, 0$; (h) no solution.

3. By definition, $\sqrt{x^2}$ is the non-negative square root of x^2. Therefore,

$$\sqrt{x^2} = \begin{cases} x, & \text{if } x \geqq 0 \\ -x, & \text{if } x < 0. \end{cases}$$

This is also the definition of $|x|$.

1-4 The Distance between Two Points on a Number Line

1. (a) $-8, 12$; (b) $-11, -1$; (c) 3; (d) no solution; (e) $8, 12$; (f) $-12, -8$; (g) $4, 8$; (h) 6.

3. (a) 4; (b) $\frac{7}{2}$; (c) $(c + d)/2$.

5. (a) 7; (b) 2; (c) $\frac{98}{11}$; (d) $\dfrac{10q - 2}{1 + q}$.

7. (a) 120; (b) $d(t) = |t^2 + 2t|$.

1-6 Some Subsets of a Plane

1.

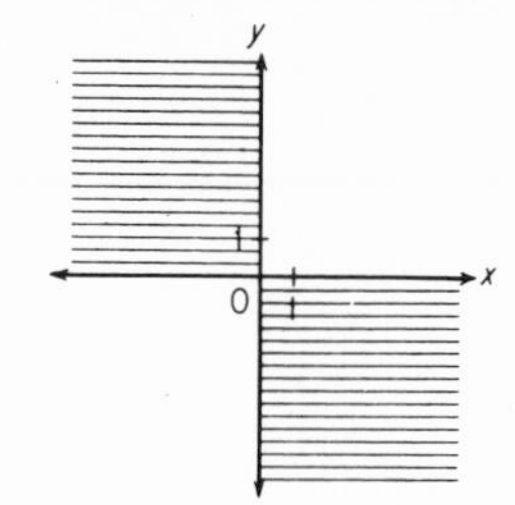

3.

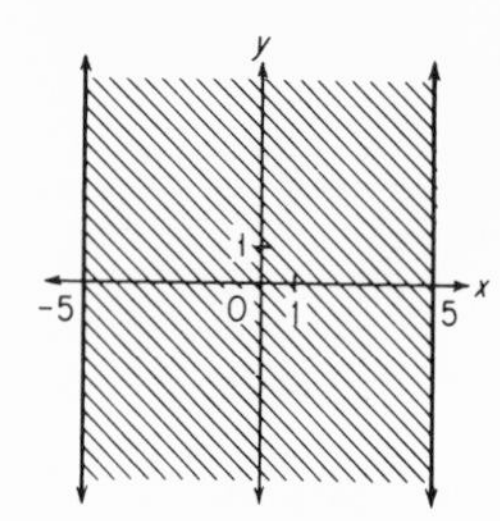

5.

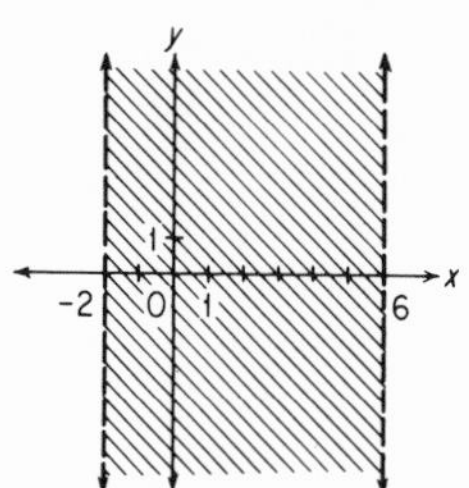

7.

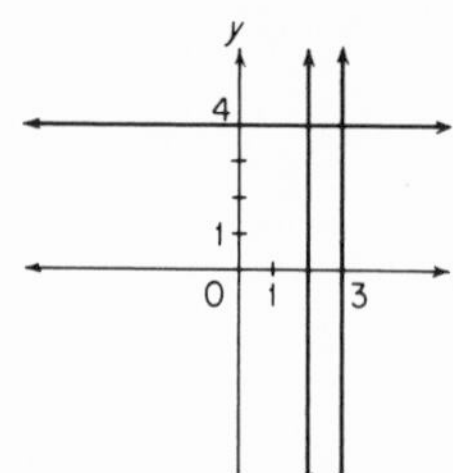

9.

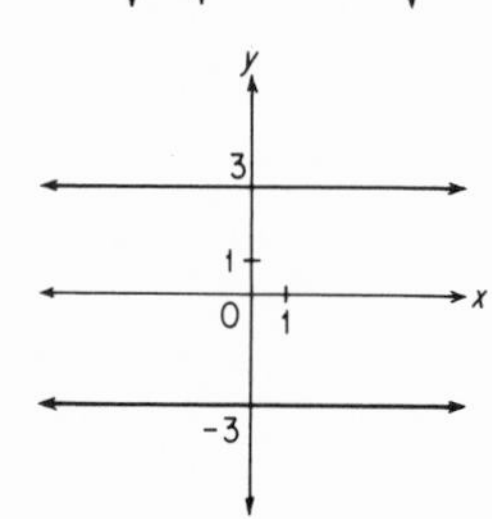

11.

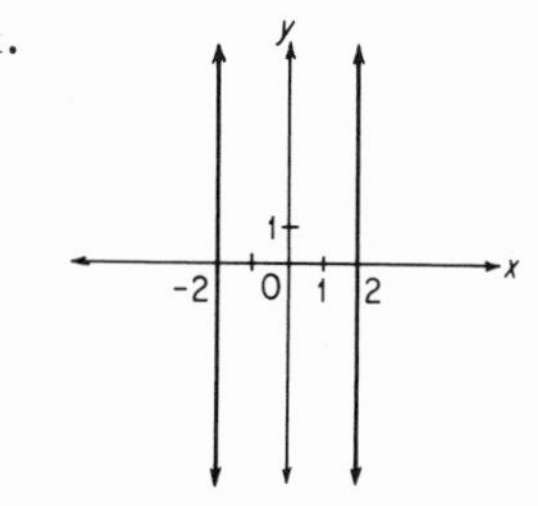

1-7 The Line Determined by Two Points in a Coordinate Plane

1. (a) The graph of $\{(x, y): y - 5 = -2(x - 12)\}$.
(b) The graph of $\{(x, y): y = x\}$.
(c) The graph of $\{(x, y): y = -x\}$.
(d) The graph of $\{(x, y): y - 3 = 0\}$.
(e) The graph of $\{(x, y): x = 5\}$.
(f) The graph of $\{(x, y): y - 4 = -(x - 3)\}$.
(g) The graph of

$$\{(x, y): y + 3 = \frac{-11}{2}(x + 2)\}.$$

Each answer has more than one correct form.

3. (a) No. The line that contains (0, 0) and (2, 3) has equation $y = (\frac{3}{2})x$. The point (4, 5) does not satisfy this.
(b) No. The line that contains $(-1, 2)$ and (3, 4) has equation $y - 2 = \frac{1}{2}(x + 1)$. The point (100, 50) does not satisfy this.
(c) Yes. The line that contains (1, 4) and (3, 7) has equation $y - 4 = (\frac{3}{2})(x - 1)$. The point (5, 10) satisfies this.

5. Since (p, q) and (r, s) satisfy the equation of the line: $s - b = m(r - a)$ and $q - b = m(p - a)$. This implies (after subtraction) that $s - q = m(r - p)$, which is equivalent to the desired condition.

1-8 Subsets Described by $Ax + By = C$

1.

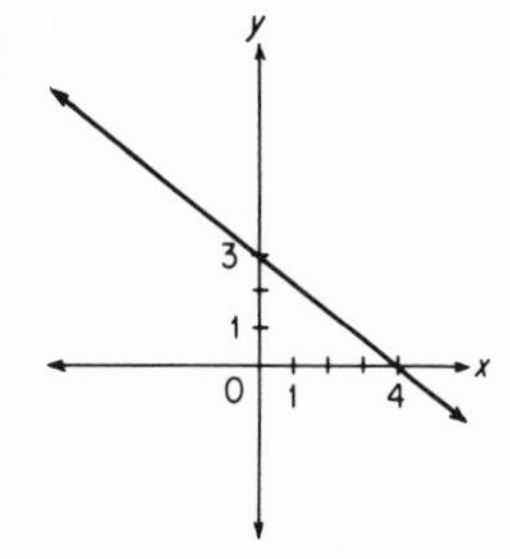

3.

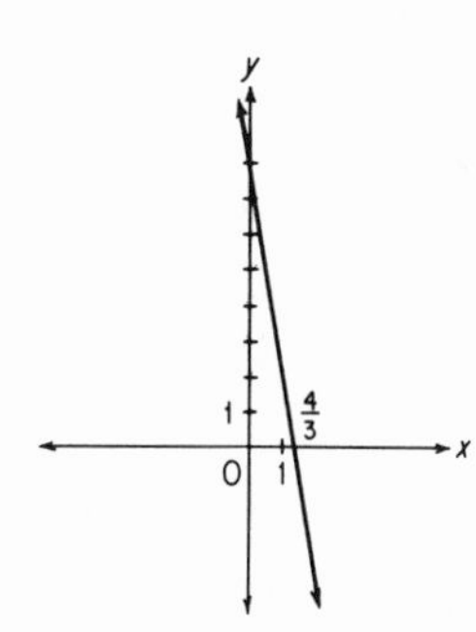

5.

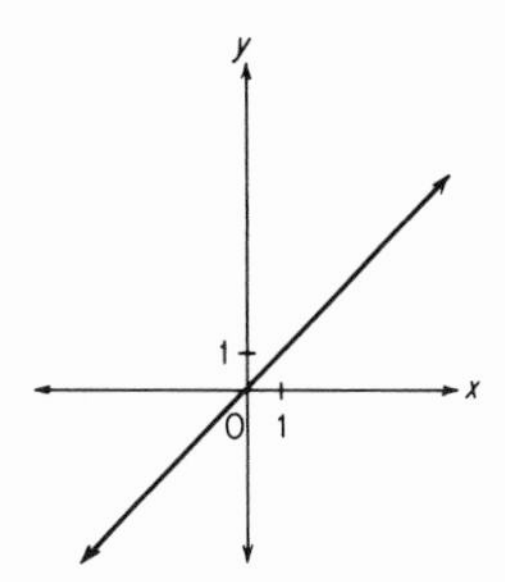

7.

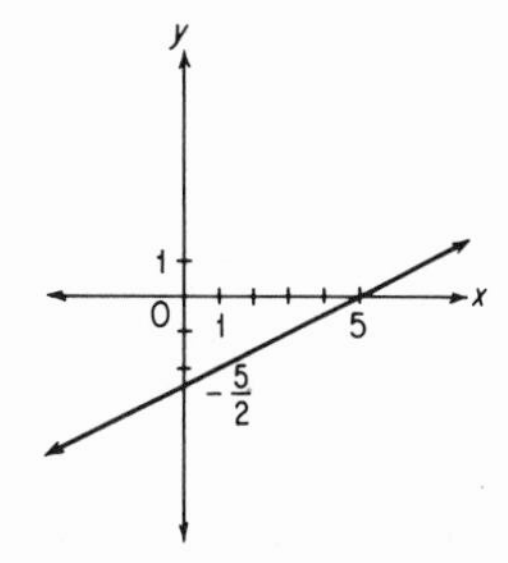

9.

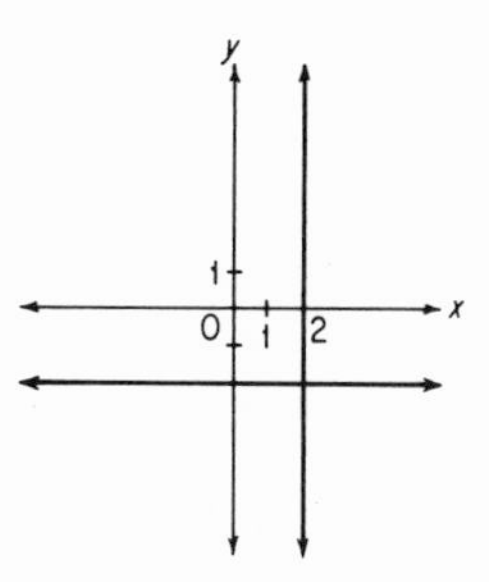

11.

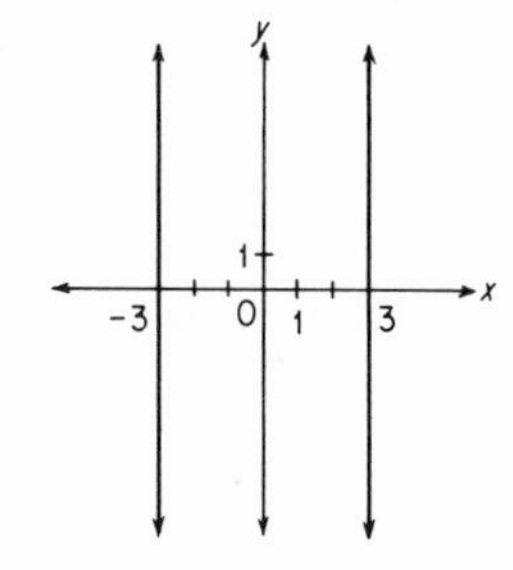

13. The graph is the single point that is the intersection of two lines.

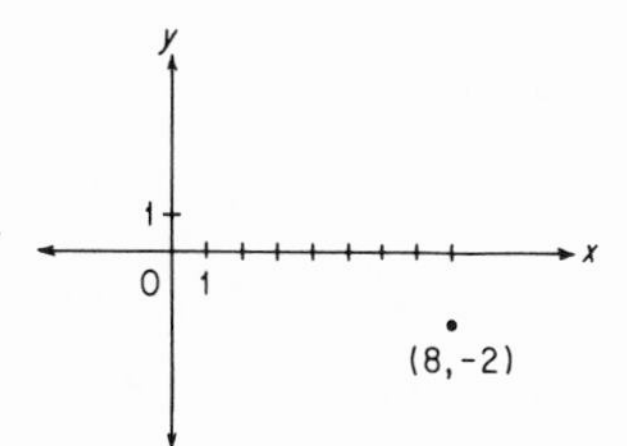

15. The set is empty.

1-9 Half-Planes

1.

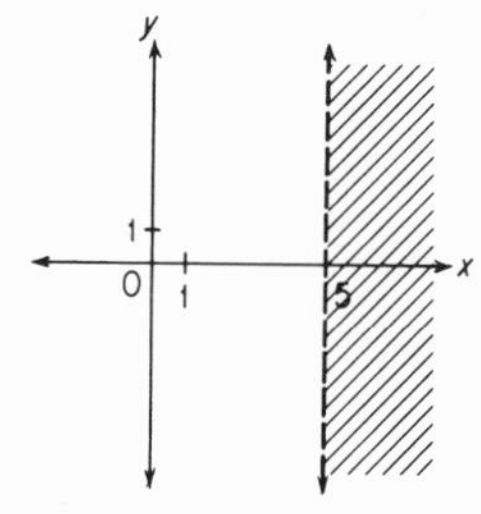

3.

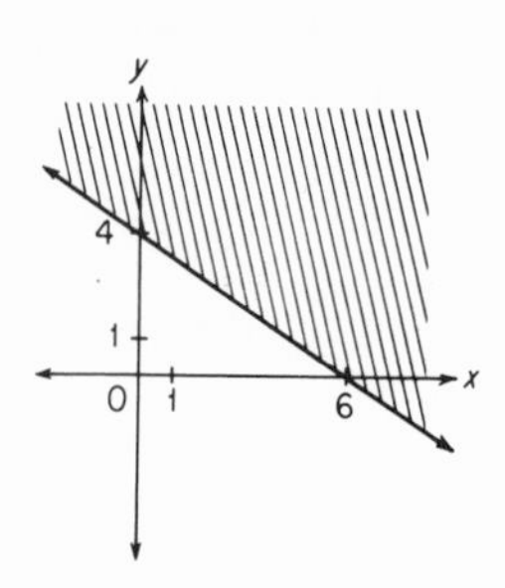

5.

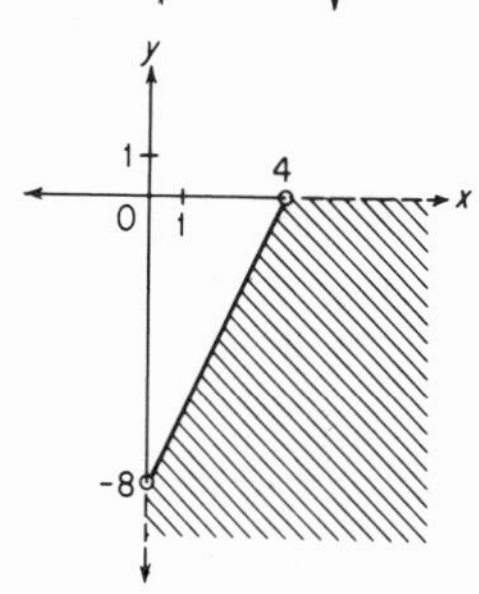

1-10 Families of Lines

1. (a) $\{(x, y): y + 2 = m(x - 4)\}$.

   (b) $\{(x, y): y + 2 = \frac{-1}{2}(x - 4)\}$.

   (c) $\{(x, y): y + 2 = -2(x - 4)\}$.

3.

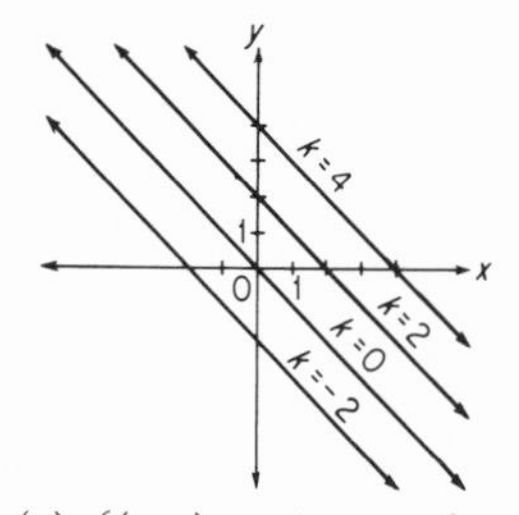

(a) $\{(x, y): x + y = 0\}$.
(b) $\{(x, y): x + y = -71\}$.

1-11 The Distance between Two Points in a Coordinate Plane

1. (a) 7; (b) 4; (c) $6\sqrt{2}$; (d) 5; (e) $\sqrt{53}$; (f) $25\sqrt{17}$.

3. (a) 10; (b) 8.

5. Use of the distance formula shows the squares of the lengths of the three sides of this triangle to be 64, 36, and 100. Since $100 = 64 + 36$, the triangle is a right triangle.

7. (a) $(1, 3)$; $(-9, -2)$.
   (b) $d(t) = \sqrt{(1 - 2t)^2 + (3 - t)^2}$.
   (c) $d(t) = \sqrt{(1 + 2t)^2 + t^2}$.
   (d)

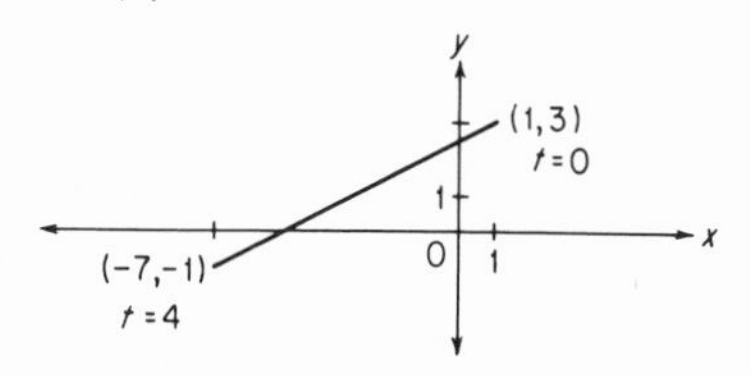

1-12 Circles and Disks

1. (a) $\{(x, y): x^2 + y^2 = 1\}$; (b) $\{(x, y): (x - 3)^2 + (y - 5)^2 = 1\}$.

3.

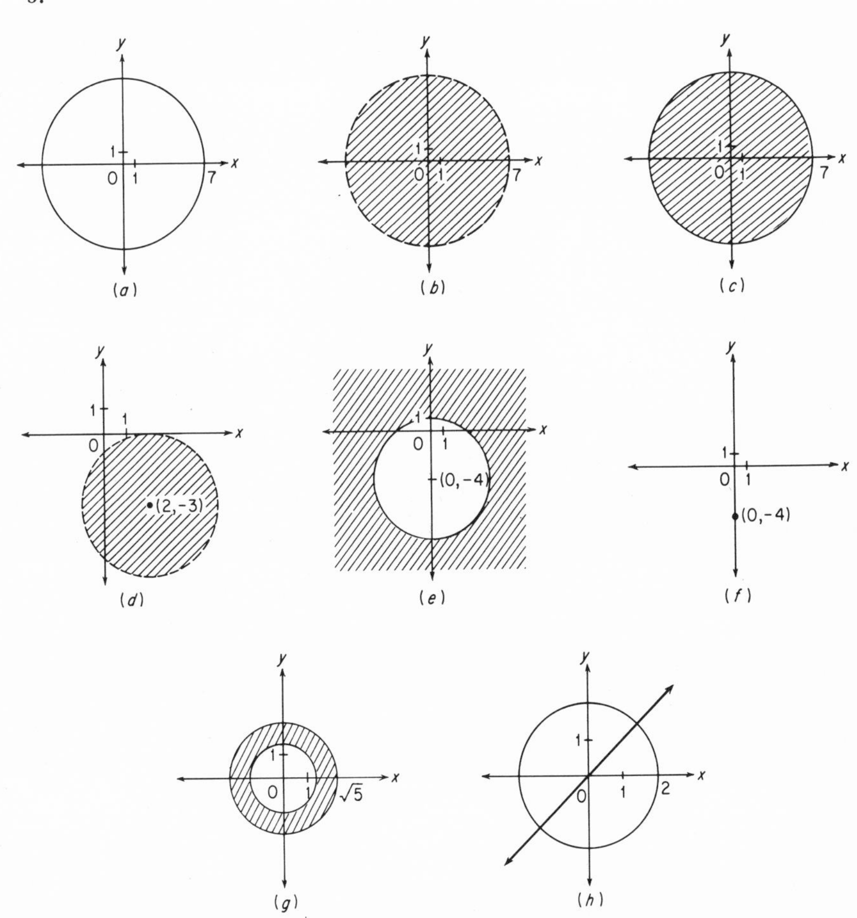

5. $\{(x, y): 4 < x^2 + y^2 < 6\}$.

CHAPTER 2—THE IDEA OF A FUNCTION AND SIMPLE EXAMPLES

2-1 What Is a Function?

1. (a) 15; (b) 20; (c) 50; (d) 7; (e) 2; (f) 20; (g) No; (h) $\{5, 10, 15, 20, \ldots\}$.

3. (a) 2; (b) 3; (c) $\{0, 1, 2, 3, 4\}$; (d) -1, 4, 9, and others.

5. (a) $P(S) = 200 + 0.05S$; (b) $S = \$16{,}000$; (c) $S \geqq \$16{,}000$.

7. (a) 0; (b) 3; (c) $n \geqq 31$.

9. Yes; $f = g$ because f and g have the same domain and $|x| = \sqrt{x^2}$ for all x.

2-2 The Graph of a Function

1.

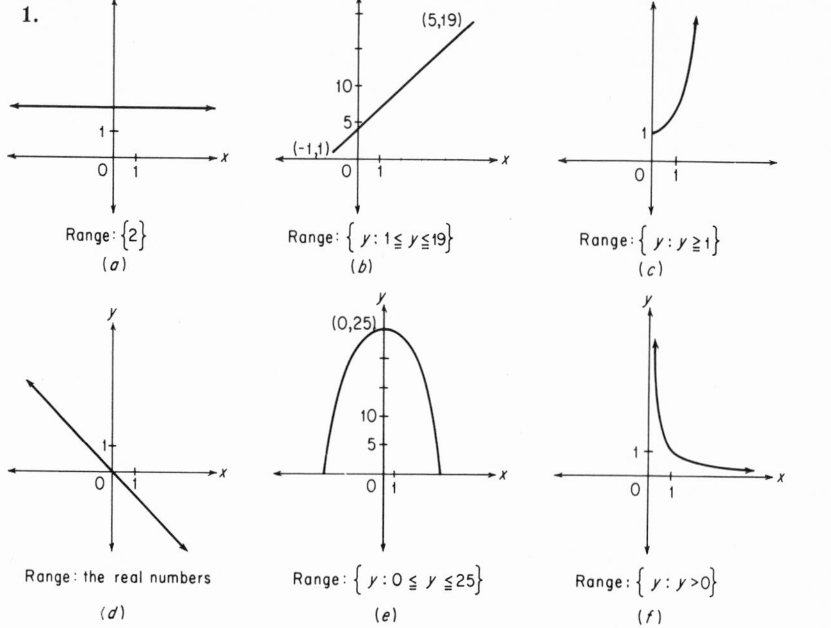

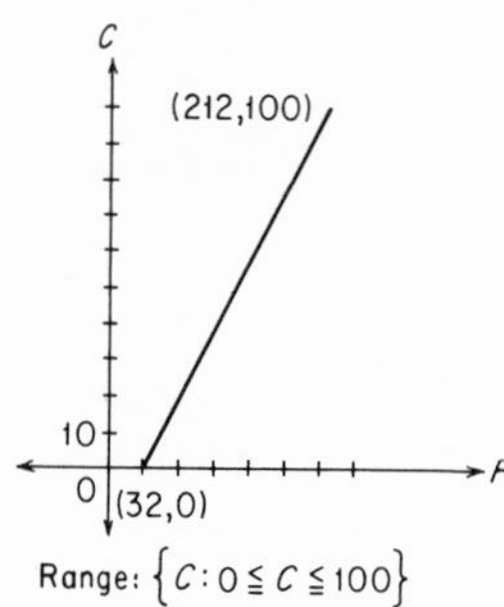

Range: $\{C: 0 \leq C \leq 100\}$

(g)

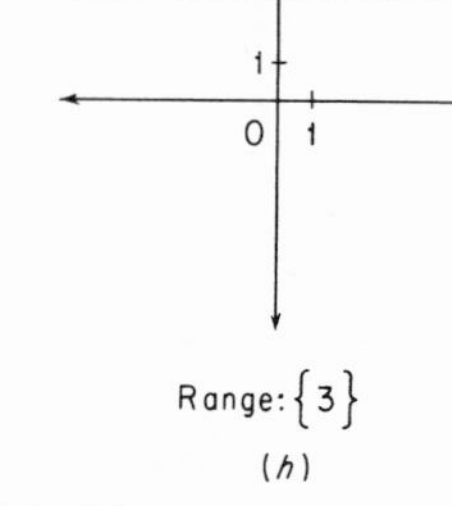

Range: $\{3\}$

(h)

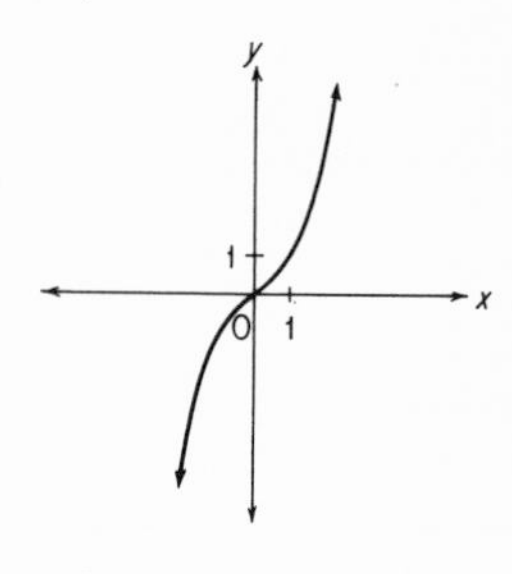

Range: the real numbers

(i)

3. The graphs of parts (b), (d), (h), and (j) are the graphs of functions. In each of the other graphs there is at least one point of the domain that has two or more images.

2-3 More on the Way a Function Is Defined

1. $\{x: x \neq -2\}$.

3. $\{x: x \neq 2\}$.

5. $\{x: x \neq 2 \text{ and } x \neq -2\}$.

7. The real numbers.

2-4 The Zeros of a Function

1. (a) $-\frac{4}{3}$; (b) No zeros; (c) No zeros; (d) $\sqrt{5}$, $-\sqrt{5}$; (e) 7, 4; (f) 0, 3; (g) -1; (h) No zeros; (i) -1, 1; (j) -4.

2-5 Functions Defined by a Multi-Part Rule

1. \$200; \$650.

3. (a) $\{w: 0 < w \leqq 320\}$.
   (b)

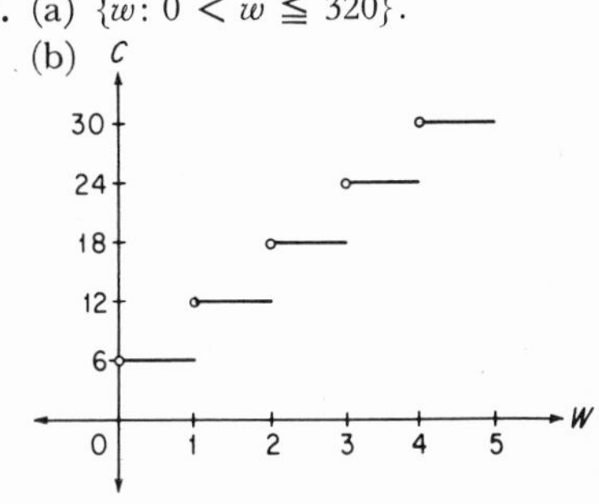

2-6 The Cartesian Product of Two Sets: Relations

1. $\{(1, 5), (1, 6), (2, 5), (2, 6), (3, 5), (3, 6)\}$.

3. (a) 8; (b) pq.

2-7 Functions from A into B

1. (a) $f = \{(1, a), (2, a), (3, a)\}$
   $g = \{(1, b), (2, a), (3, c)\}$
   There are other examples.

   (b) $\{(1, a), (1, b)\ (2, a), (3, a)\ (3, b)\}$
   $\{(1, c), (2, a), (2, b), (3, a)\}$

3.

This is a function from R into R
(a)

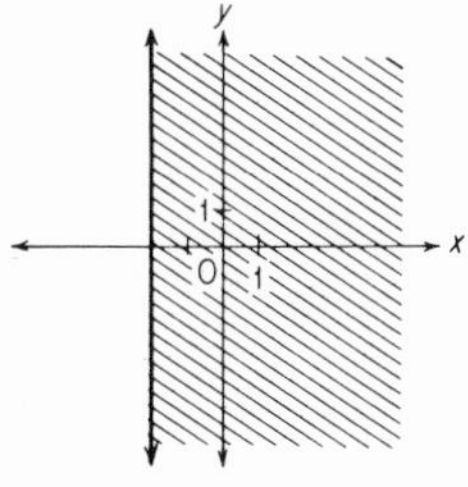

Not a function
(b)

3. (cont.)

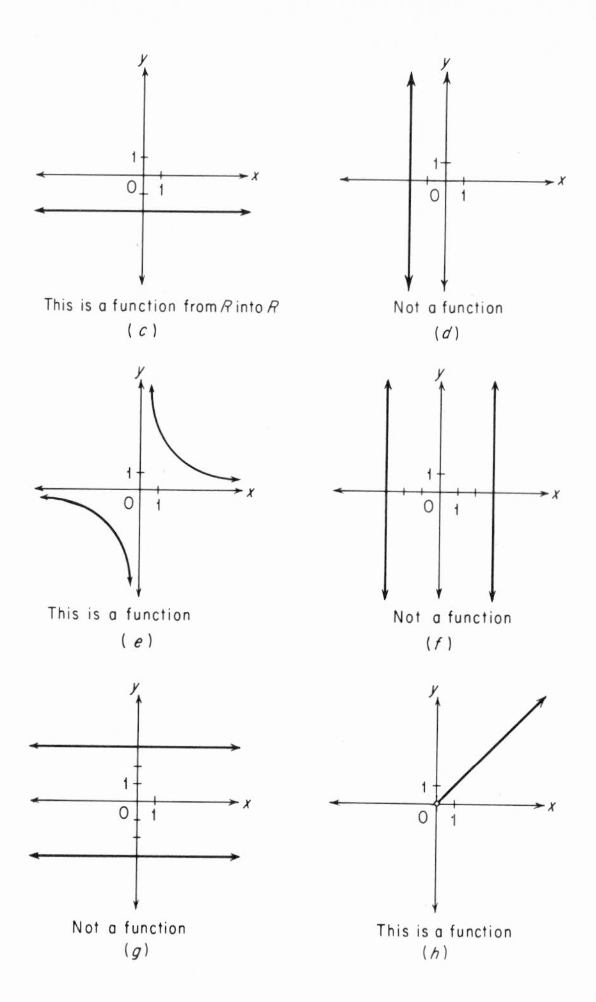

CHAPTER 3—LINEAR FUNCTIONS

3-1 Definition and Examples

1. Linear functions are defined in (a), (d), (f), (g), and (h).

3. (a) 108; (b) 120; (c) 25.

3-2 Graphs of Linear Functions

1.

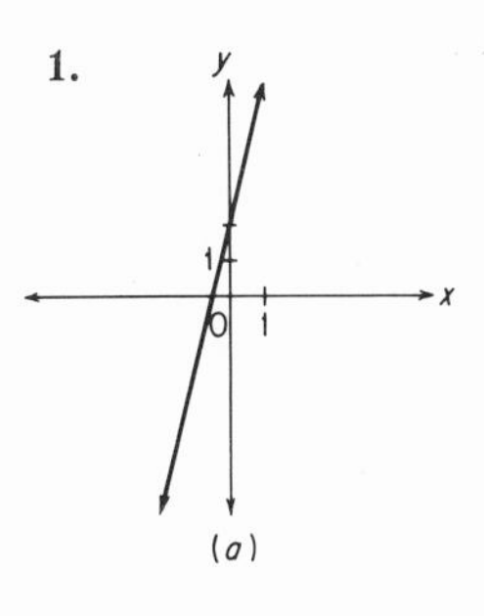

(*a*)

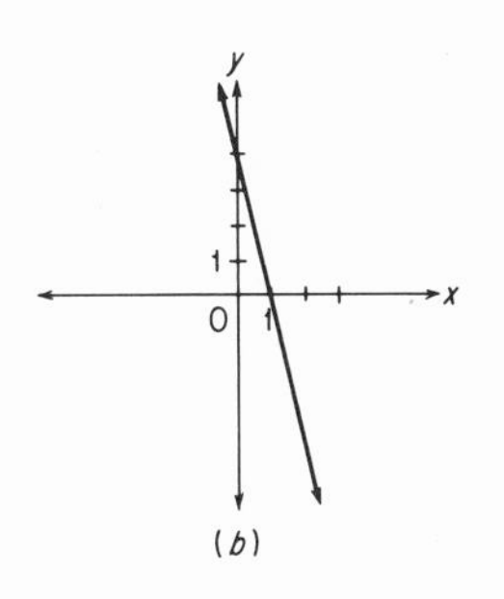

(*b*)

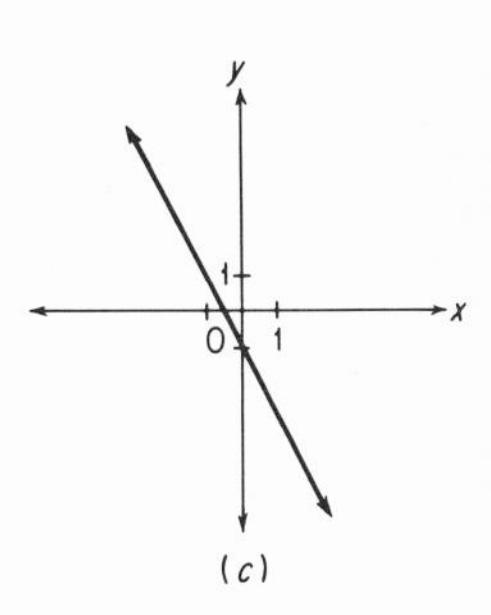

(*c*)

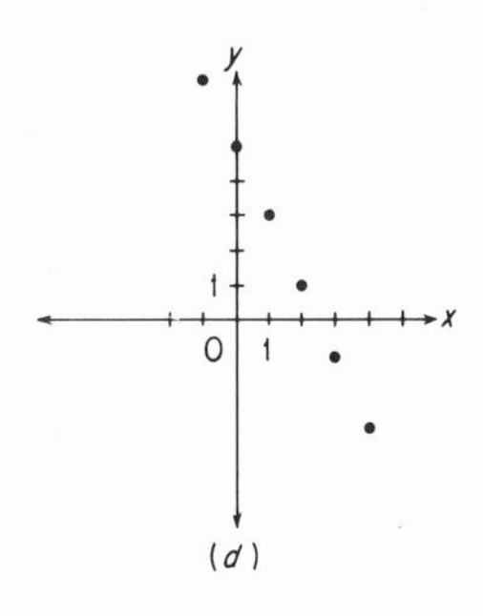

(*d*)

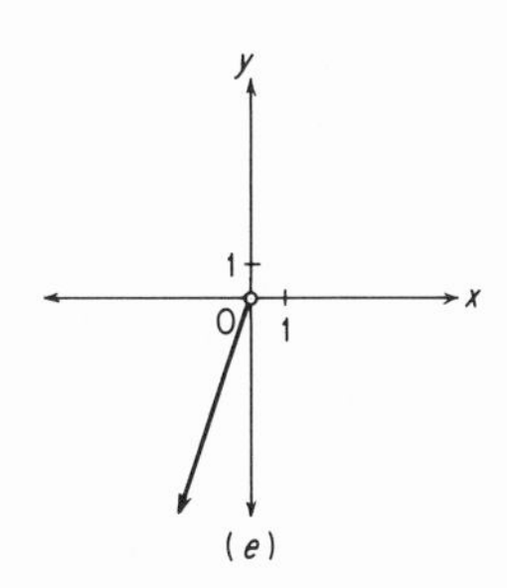

(*e*)

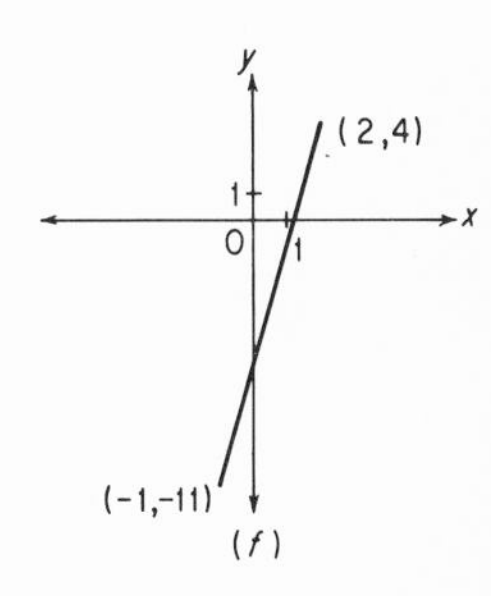

(*f*)

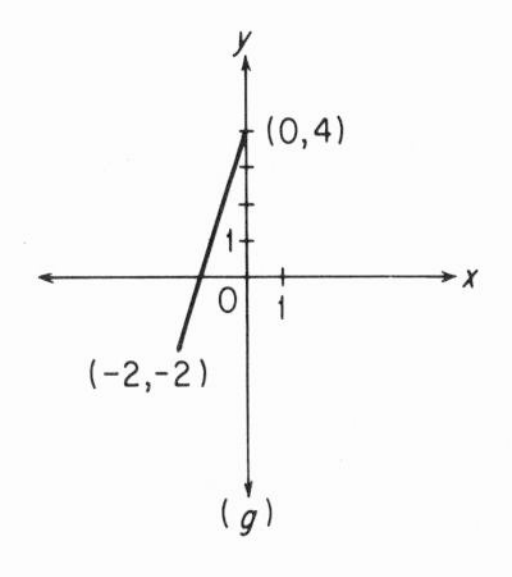

(*g*)

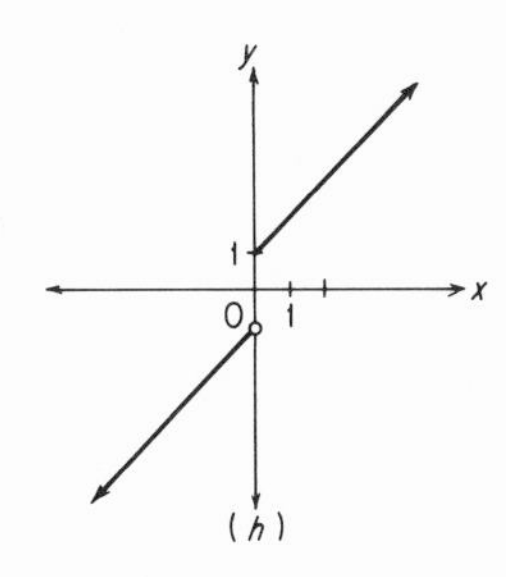

(*h*)

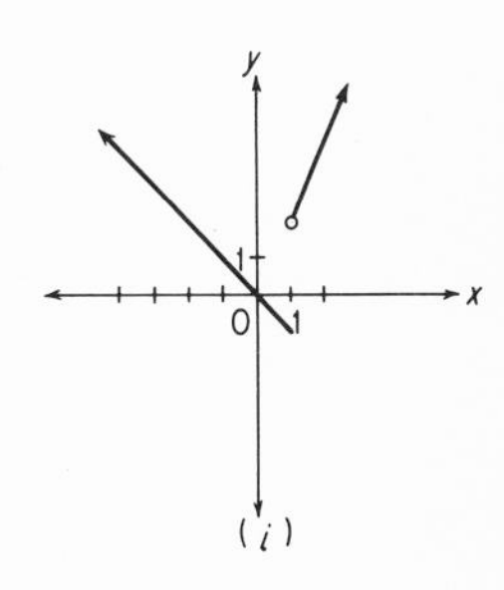

(*i*)

3-3 The Linear Function that Maps a onto b and c onto d

1. (a) 208; (b) $f(x) = 8 + 5x$.

3. $h(x) = \frac{1}{2}(9 + x)$.

5. (a) 121; (b) $f(n) = 4 + 3(n - 1)$; (c) A sequence of numbers $a_1, a_2, a_3, \ldots, a_n, \ldots$ is called an arithmetic progression if and only if $a_n - a_{n-1}$ is the same number d (called the common difference) for all $n \geqq 2$. Suppose $f(x) = mx + b$, domain: the positive integers. Then

$$f(n) - f(n - 1) = mn + b - [m(n - 1) + b] = m$$

Therefore, the set of images that f generates is an arithmetic progression with the common difference equal to m.

3-4 The Solution of Linear Equations in One Unknown

1. (a) $-\frac{2}{3}$; (b) $-\frac{20}{3}$; (c) 800; (d) No solution; (e) $-\frac{6}{5}$; (f) $\frac{1}{5}$; (g) 0; (h) $\frac{80}{11}$; (i) $5/(\sqrt{2} + 1)$; (j) 0; (k) $\frac{5}{7}$; (l) 2.

3. (a) 28; (b) 52; (c) 4; (d) 32; (e) 8.

5. Yes; $-40°$F is $-40°$C.

7. (a) $f(x) = x + 1$ has no fixed point; (b) $m \neq 1$.

3-5 Some Special Properties of Linear Functions

1. (a) $\frac{7}{5}$; (b) $-\frac{2}{7}$; (c) $\frac{3}{4}$; (d) $-3/\sqrt{2}$; (e) -14; (f) 5; (g) $\pi/3$; (h) 5.

3. (a) $f(x) = x^2$ and other examples;

(b) $g(x) = \begin{cases} 1, & \text{if } x \text{ is a rational number} \\ 0, & \text{if } x \text{ is an irrational number} \end{cases}$

and other examples.

3-6 Linear Interpolation

1. (a) 110; (b) 105; (c) 117.5; (d) 90.

3. (a) 1.0207; (b) 1.0041; (c) 1.0754; (d) 1.0896.

5. (a) 0.7888; (b) 0.3398; (c) 0.1741; (d) 0.7146; (e) 0.6382; (f) 4.109.

CHAPTER 4—A FAMILY OF ABSOLUTE VALUE FUNCTIONS

4-1 Reflection in a Line: Symmetry

1. (a) (2, 5); (b) (−2, −5); (c) (4, −5); (d) (2, 13).

3. (a) The given line and each line perpendicular to it are axes of symmetry. (b) The perpendicular bisector of a line segment is an axis of symmetry of the segment; also the line that contains the given line segment.

5.

$f(x) = 2x + 3$

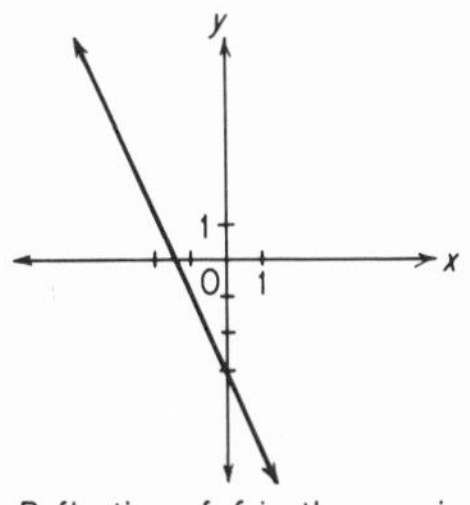

Reflection of f in the x-axis

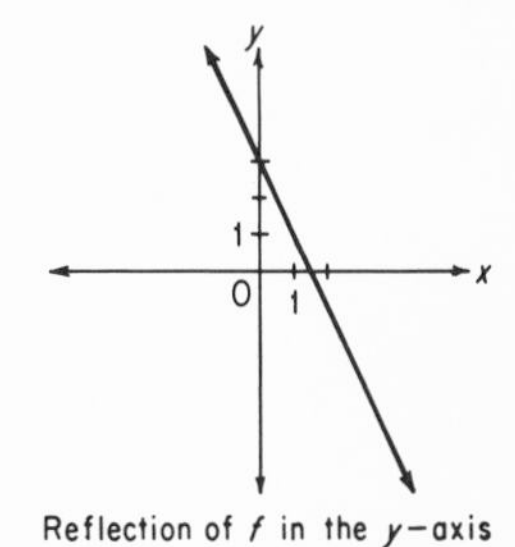

Reflection of f in the y-axis

7.

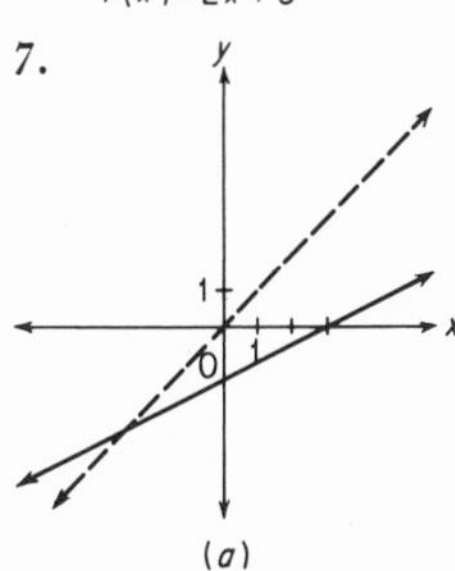

(a)

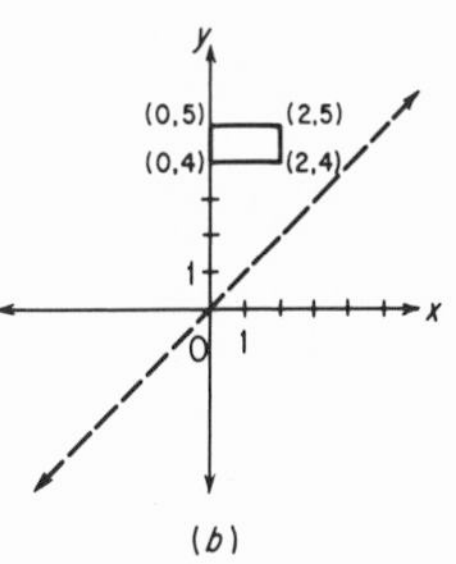

(b)

4-2 The Basic Absolute Value Function

1. (a) Neither maximum nor minimum. (b) The minimum is −2; the maximum is 0. (c) Neither maximum nor minimum. (d) No minimum; the maximum is 5.

3. Minimum is −10; maximum is 2.

5. (a) 6, −6; (b) No solution; (c) −4, 8; (d) −8, 4; (e) −2, 4; (f) 3; (g) 3; (h) $(10 + \pi)/2$.

4-3 Some Variations of the Basic Absolute Value Function

1. Let x denote an arbitrary number. Then $x = 2 + s$ or $x = 2 - s$ for a unique number $s \geqq 0$. $F(2 + s) = |s|$; also $F(2 - s) = |-s| = |s|$. The graph of F contains the point $(2 + s, |s|)$ and also the point $(2 - s, |s|)$. Each of these points is the reflection of the other in the line $x = 2$.

3. (a)

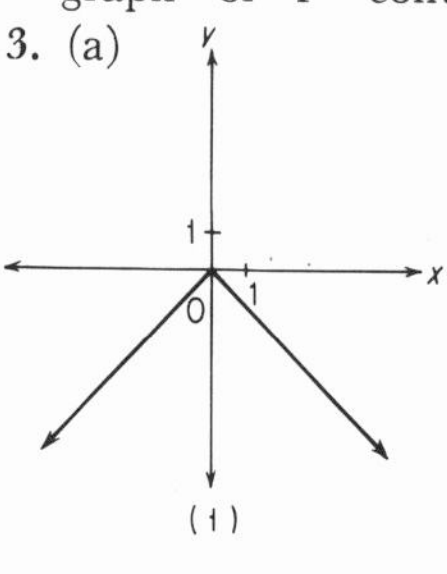

(1)

(2)

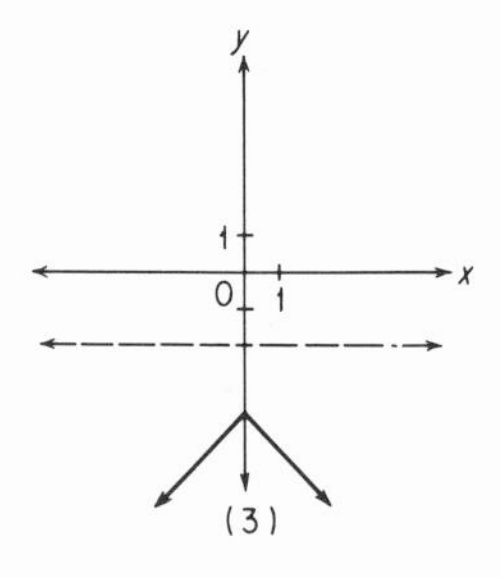

(3)

(b) $y = -|x|$, $y = |x - 8|$ $y = -|x| - 4$

4-4 A Three-Parameter Family of Functions

1. Each real number x can be written in the form $x = h + s$ or $x = h - s$, where $s \geqq 0$. Also $f(h + s) = a|s| + k$, and $f(h - s) = a|-s| + k = a|s| + k$. Therefore, for every s, $f(h + s) = f(h - s)$. The graph of f contains the points $(h + s, f(h + s))$ and $(h - s, f(h + s))$. Each of these points is the reflection of the other in the line $x = h$.

3.

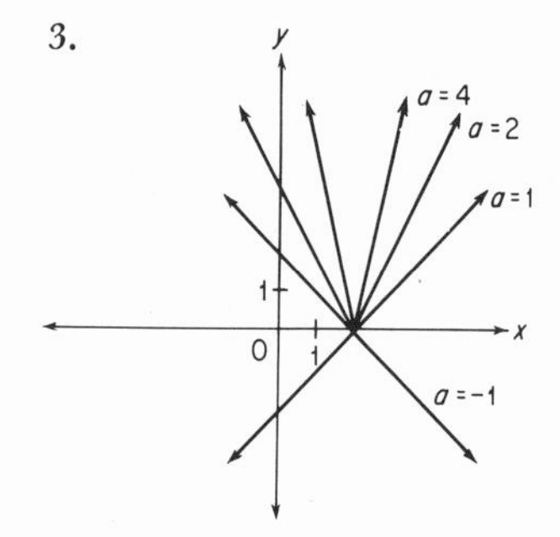

5. (a) $f(x) = |x - 3| + 5$.
(b) $f(x) = 2|x| - 2$.
(c) $f(x) = -\frac{1}{2}|x - 6|$.
(d) $f(x) = 2|x - 2| + 8$.
(e) $f(x) = -2|x + 2| + 12$.

7. (a)

(b) No; the maximum number in the range is -1.
(c) $\{-2, 0\}$.

9. (a)

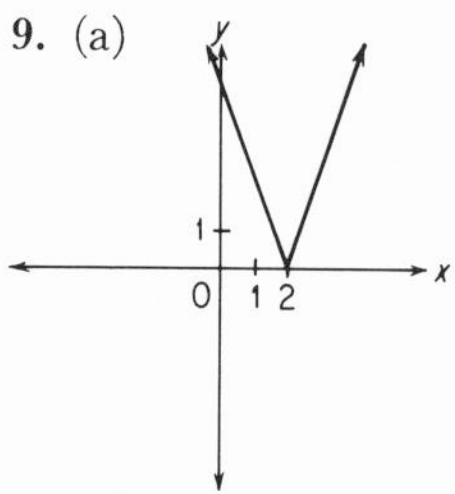

(b) Yes; 2 is the only zero.
(c) All numbers except 2.

11. Reflect the graph of f in the line $x = -1$; then reflect this (new) graph in the line $y = 6$.

CHAPTER 5—QUADRATIC FUNCTIONS

5-1 Definition and Examples

1. $y = x^2 + 6x + 9$; $a = 1$, $b = 6$, $c = 9$.

3. $y = 2x^2 + 12x + 25$; $a = 2, b = 12$, $c = 25$.

5. $y = 2x^2 + 6x + 5$; $a = 2$, $b = 6$, $c = 5$.

7. $y = x^2 + 4x - 5$; $a = 1$, $b = 4$, $c = -5$.

9. $y = x^2 + 2x - 8$; $a = 1$, $b = 2$, $c = -8$.

5-2 A Review of Some Facts about Numbers

1. (a) $y = x^2 + x - 6$.
(b) $y = -4x^2 - 8x - 4$.
(c) $y = -x^2 + 10x - 25$.
(d) $y = 3x^2 + 12x + 12$.
(e) $y = x^2 - 9$.
(f) $y = 4x^2 - 100$.
(g) $y = -x^2 - x + 6$.
(h) $y = -x^2 - x + 20$.
(i) $y = 10x^2 - 10\pi^2$.
(j) $y = \pi x^2 + 14\pi x + 49\pi$.
(k) $y = x^2 + 4x$.
(l) $y = 3x^2 - 15x$.
(m) $y = 7x^2 + 7x - 42$.
(n) $y = -2x^2 - 2x + 24$.

3. (a) -5; (b) 4, -1; (c) -1, $-\frac{1}{2}$; (d) 0, -4; (e) 0, $\frac{1}{7}$; (f) 5, -5; (g) 5, -5, (h) 10, -10; (i) 1, -6; (j) 0, $\frac{8}{3}$.

5-3 Another Form for the Rule of a Quadratic Function

1. (a) $y = x^2 + 6x + 4$.
(b) $y = 2x^2 - 4x + 2$.
(c) $y = x^2 + 14x + 59$.
(d) $y = -4x^2 - 16x - 9$.
(e) $y = x^2 - 2x + 4$.
(f) $y = 2x^2 + 6x + 5$.
(g) $y = 5x^2 - 4x + 1$.
(h) $y = 7x^2 + 42x + 23$.
(i) $y = x^2 - 8x$.
(j) $y = 3x^2 - 14x + 21$.

5-4 Graphing Quadratic Functions

1.

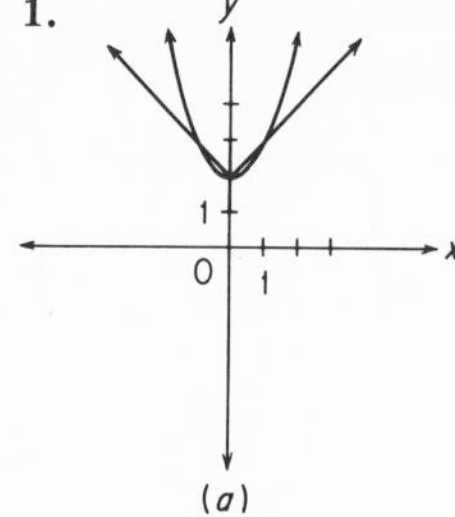

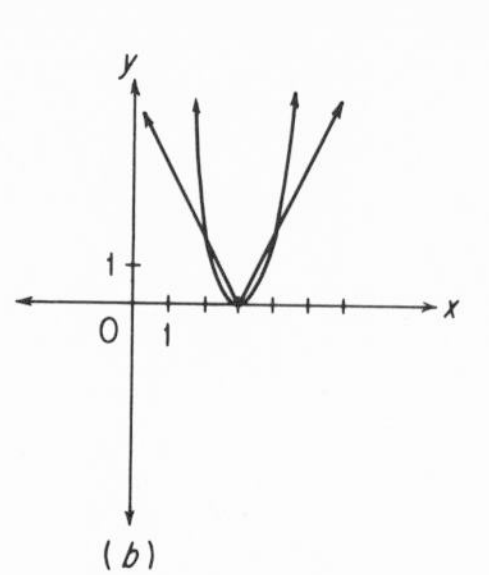

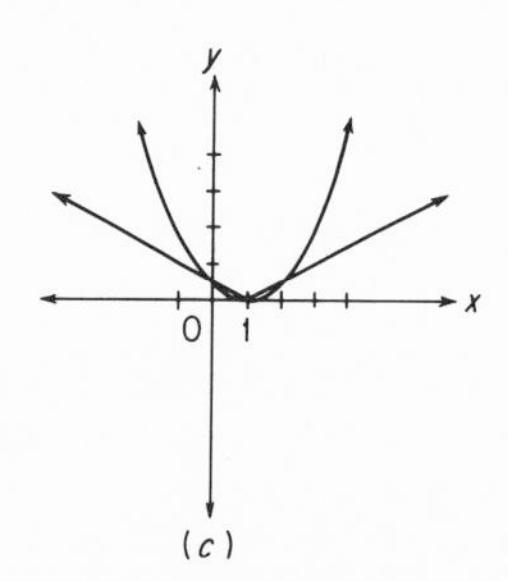

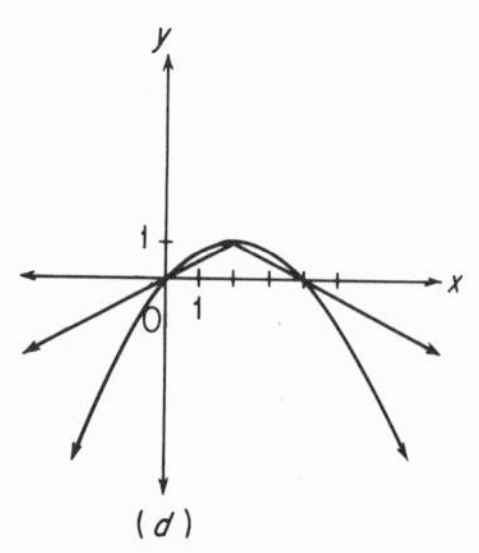

3. (a) min.: 5; (b) min.: 5; (c) min.: -4; (d) max.: -4; (e) min.: -9; (f) min.: $\frac{3}{4}$; (g) max.: 0; (h) min.: -4; (i) min.: -7; (j) min.: 0.

5.

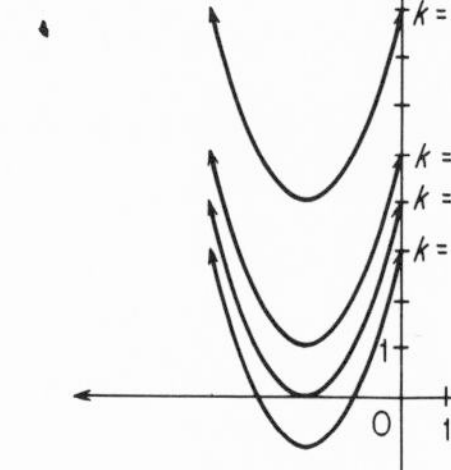

7. (a) $y = 4(x - 2)^2 + 3$ or $y = -4(x - 2)^2 + 3$.
(b) $y = 5(x + 1)^2 + 6$ or $y = -5(x + 1)^2 + 6$.
(c) $y = 2x^2$.
(d) $y = \dfrac{-5}{4}(x - 2)^2 + 5$.

9. (a)

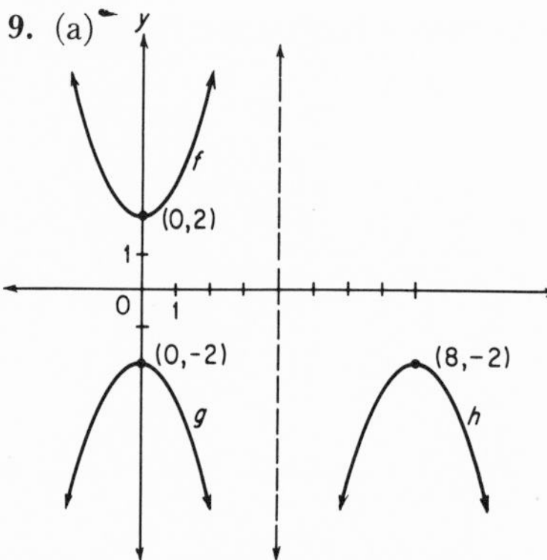

(b) $g(x) = -x^2 - 2$;
$h(x) = -(x - 8)^2 - 2$.

5-5 Does the Quadratic Function f Map any Number onto c?

1. 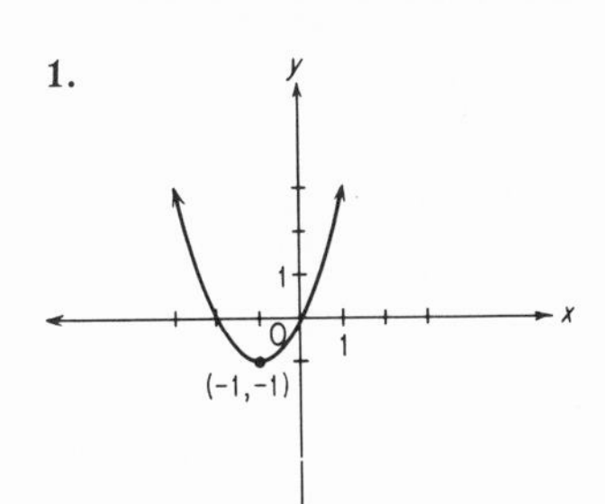

(a) 0, -2; (b) -3, 1; (c) -1; (d) None.

3. (a) 8, -1; (b) -4; (c) 0, $\frac{1}{5}$; (d) $\sqrt{5}$, $-\sqrt{5}$; (e) 5, -1; (f) 1, $-\frac{1}{2}$.

5-6 Does the Quadratic Function f Have any Zeros? If So, What Are They?

1. (a) 0, 6; (b) $(1 \pm \sqrt{13})/6$; (c) ± 2; (d) None; (e) $(-1 \pm \sqrt{5})/2$; (f) None; (g) 1; (h) None; (i) 3, -1; (j) 1, $-\frac{3}{2}$; (k) None; (l) $(-3 \pm \sqrt{65})/4$.

3. 5.

5. -1.

7. $\frac{1}{2}$.

5-7 Solving Quadratic Equations

1. (a) 0, 7; (b) 0, $\frac{4}{3}$; (c) -4; (d) 1, $-\frac{1}{2}$; (e) 5, -3; (f) $(-3 \pm \sqrt{5})/2$; (g) No real number is a solution; (h) 1; (i) 0, -4; (j) -2, -3; (k) -2, 6; (l) -3, $\frac{1}{2}$.

3. $a \neq 1$ and $b^2 - 8(a - 1) > 0$.

5-8 Some Quadratic Inequalities

1. (a) 0, 9.
(b) $\{x: x < 0 \text{ or } x > 9\}$.
(c) $\{x: 0 < x < 9\}$.
(d) $\{x: -1 < x < 10\}$.
(e) $\{x: -2 < x < 11\}$.
(f) $\{x: 4 < x < 5\}$.

3. (a) 4, -3.
(b) $\{x: x < -3 \text{ or } x > 4\}$.
(c) $\{x: -3 < x < 4\}$.

5-10 Some Applications

1. (a) 4.5; (b) 324 ft; (c) $t = 9$.

5. Let x denote the width of the rectangle. Then $(P/2) - x$ denotes the length, and the rule for the function that gives the area is $A(x) = x[(P/2) - x]$. This is a quadratic function; $P/4$ is the number in the domain that produces the maximum area. This implies that for maximum area the four sides of the rectangle must be equal in length.

9. (a) The circumference of the second disk is 3 times that of the first disk. (b) The area of the second disk is 9 times that of the first disk.

3. (a) $M(v_0) = (v_0^2)/64$; (b) 240 ft/sec.

7. 7 in. $\times$ 14 in.

CHAPTER 6—USING FUNCTIONS TO CONSTRUCT OTHER FUNCTIONS

6-1 Algebraic Operations with Functions

1. (a) The set of real numbers is the domain for all these functions except f/g and g/f.
$(f + g)(x) = 3x + 3$;
$(f - g)(x) = x - 1$;
$(g - f)(x) = -x + 1$;
$(f \cdot g)(x) = (2x + 1)(x + 2)$;
$(g \cdot f)(x) = (x + 2)(2x + 1)$;
$(f/g)(x) = (2x + 1)/(x + 2)$, domain: $\{x: x \neq -2\}$;
$(g/f)(x) = (x + 2)/(2x + 1)$, domain: $\{x: x \neq -\frac{1}{2}\}$;
$7 \cdot f(x) = 7(2x + 1)$;
$(-2g)(x) = -2(x + 2)$.
(b) $(f + g)(8) = 27$; $(f/g)(3) = \frac{7}{5}$; $(f \cdot g)(-1) = -1$; $(3 \cdot f)(0) = 3$.

3. (a) Valid. $(f + g)(x) = f(x) + g(x) = g(x) + f(x) = (g + f)(x)$ (since addition of numbers is commutative)
(b) Valid. The proof is similar to the proof in part (a).
(c) Invalid. Let $f(x) = x$, $g(x) = 2x$. Then $(f - g)(x) = -x$, and $(g - f)(x) = x$.
(d) Invalid. Let $f(x) = x$, $g(x) = 2x$. Then $(f/g)(x) = \frac{1}{2}$, and $(g/f)(x) = 2$.
(e) Valid. The proof is similar to the proof in part (a).
(f) Invalid. Let $f(x) = x$, $g(x) = -x + 1$. Then $(f + g)(x) = 1$, so $(f + g)$ is not a linear function.
(g) Valid. Let $f(x) = mx + b$, where $m \neq 0$; $g(x) = px + q$, where $p \neq 0$. Then $(f \cdot g)(x) = mpx^2 + (mq + pb)x + bq$. This is a quadratic function, since $mp \neq 0$.
(h) Invalid. Let $f(x) = x^2$, $g(x) = -x^2 + 1$. Then $(f + g)(x) = 1$.

5.

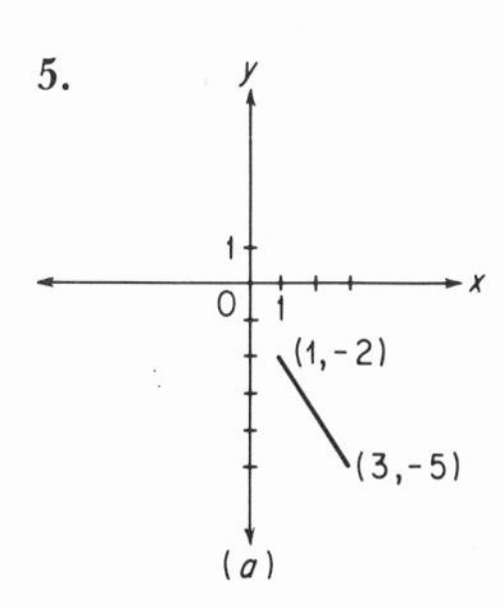

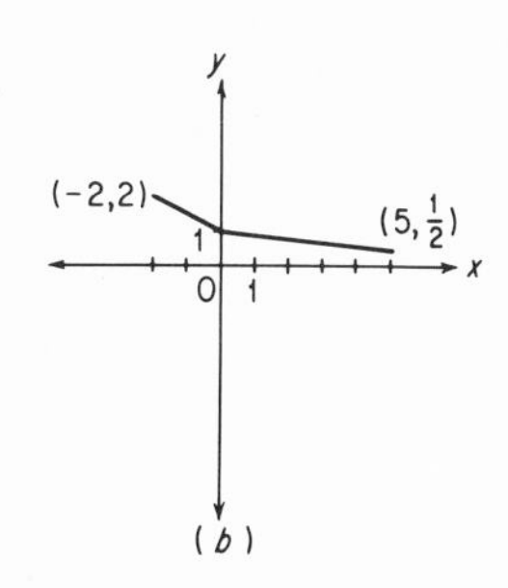

(c) The graph of $-f$ is the reflection of the graph of f in the x-axis. This is a "rigid motion." The two graphs are congruent.

6-2 Restrictions and Extensions of Functions

1.

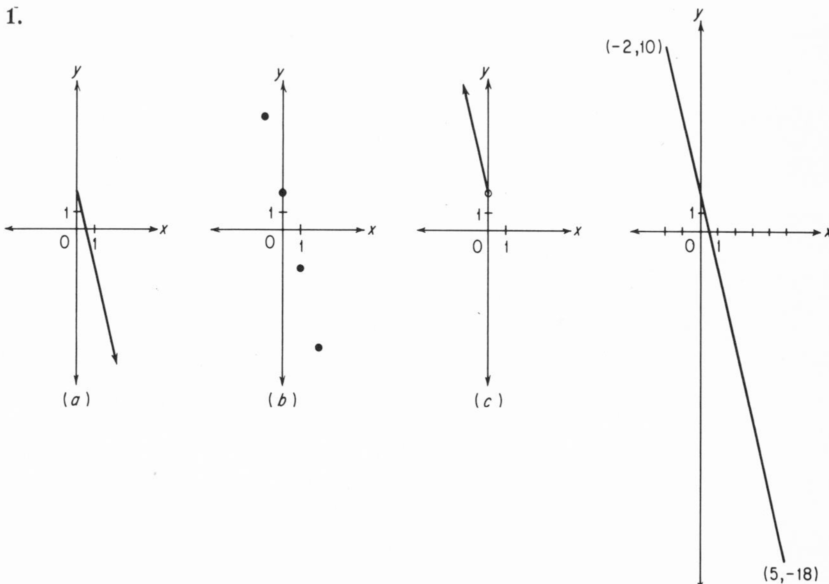

3. (a) Neither maximum nor minimum.
   (b) No maximum; minimum is 3.

5.

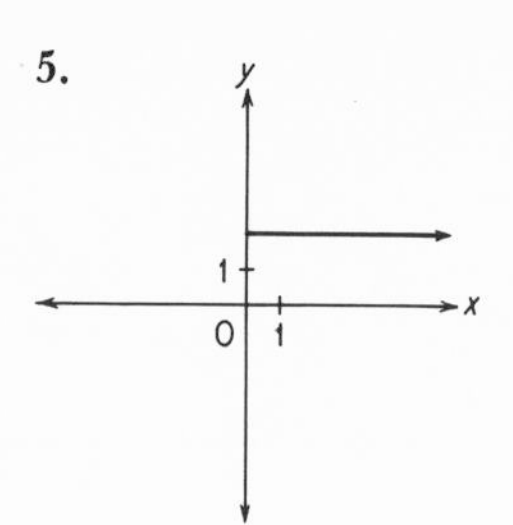

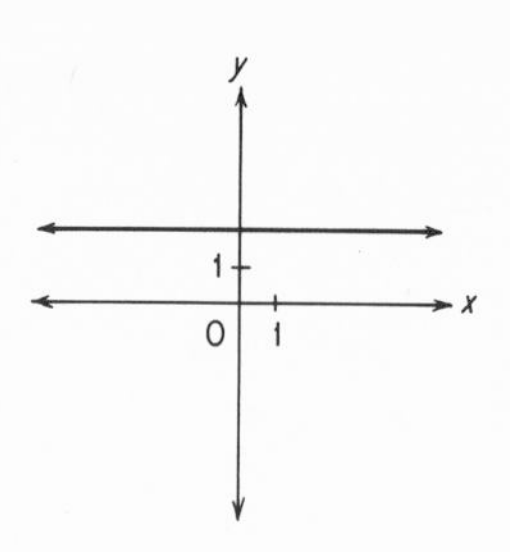

$H(x) = 2$; Domain: the real numbers

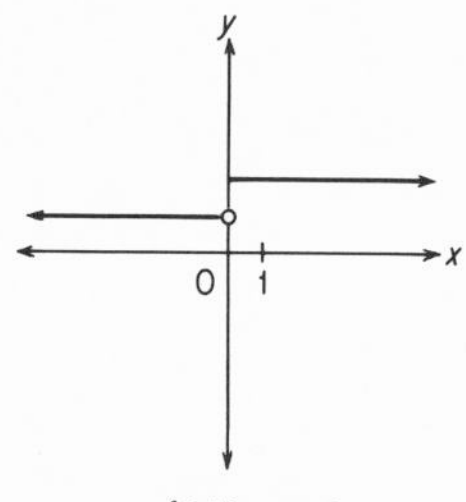

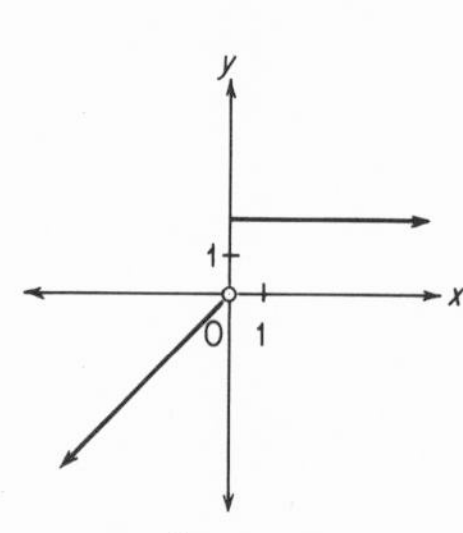

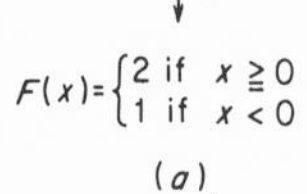

$F(x) = \begin{cases} 2 \text{ if } x \geq 0 \\ 1 \text{ if } x < 0 \end{cases}$

(*a*)

$G(x) = \begin{cases} 2 \text{ if } x \geq 0 \\ x \text{ if } x < 0 \end{cases}$

(*b*)

6-3 Composition of Functions

1. (a) -6; (b) 2; (c) $(fg)(x) = -3x$; (d) $(gf)(x) = 8 - 3x$.

5. (a) $(fg)(x) = 2/(x - 2) + 1$, domain: $\{x: x \neq 2\}$; (b) $\frac{5}{3}$.

7. Example: $f(x) = x - 5$, $g(x) = x + 5$.
Example: $f(x) = 5x + 3$, $g(x) = (\frac{1}{5})(x - 3)$.

9. (a) Let $(x, f(x))$ denote an arbitrary point on the graph of f; its reflection in the y-axis is $(-x, f(x))$. The point $(-x, f(x))$ is on the graph of fg, since $(fg)(-x) = f(x)$. Therefore, the graph of fg contains the reflection of the graph of f in the y-axis. Similarly, let $(x, (fg)(x))$ denote a point on the graph of fg; it is the reflection in the y-axis of $(-x, (fg)(x))$. This point, $(-x, (fg)(x))$, is on the graph of f since $(fg)(x) = f(-x)$. Therefore, every point on the graph of fg

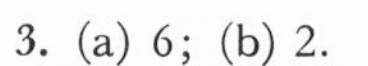

3. (a) 6; (b) 2.
(c) $(fg)(x) = |x - 2|$

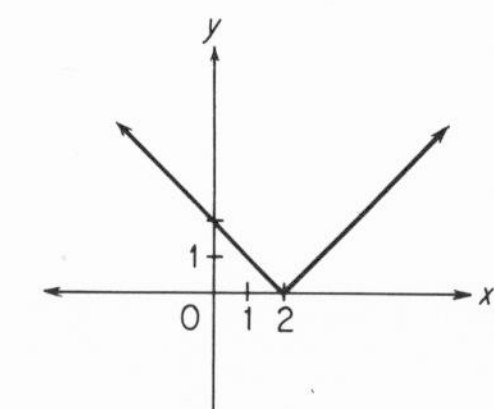

(d) $(gf)(x) = |\,|x| - 2|$

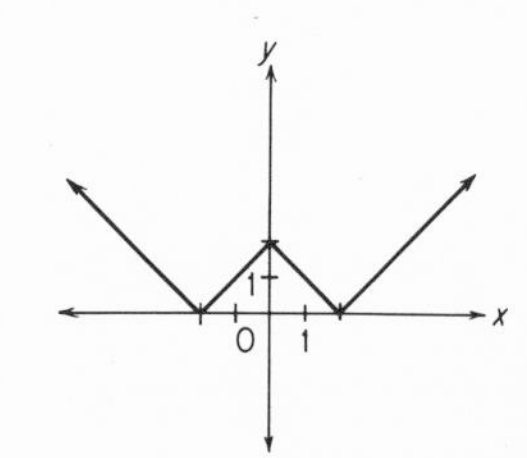

is the reflection of a point on the graph of f.
(b) Yes; reflection in an axis is a "rigid motion."
(c)

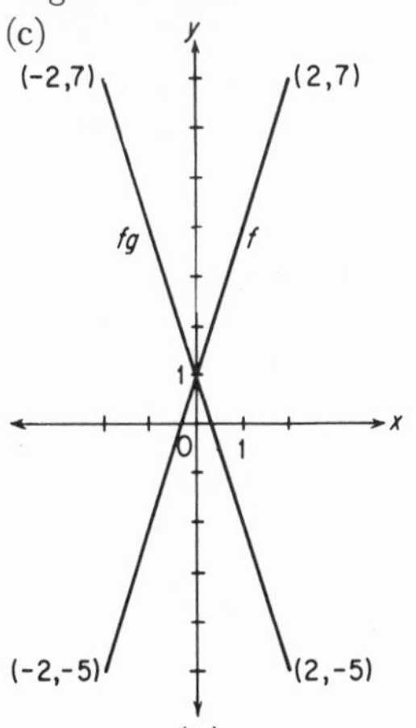

(c)

(d)

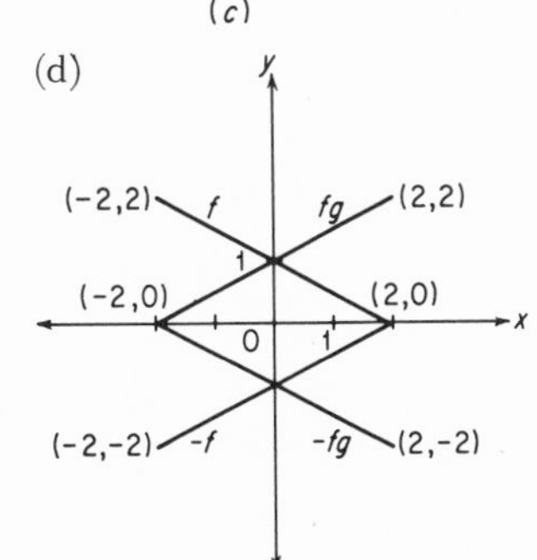

11. (a) $(fg)(x) = |x + 3|$.
(b)

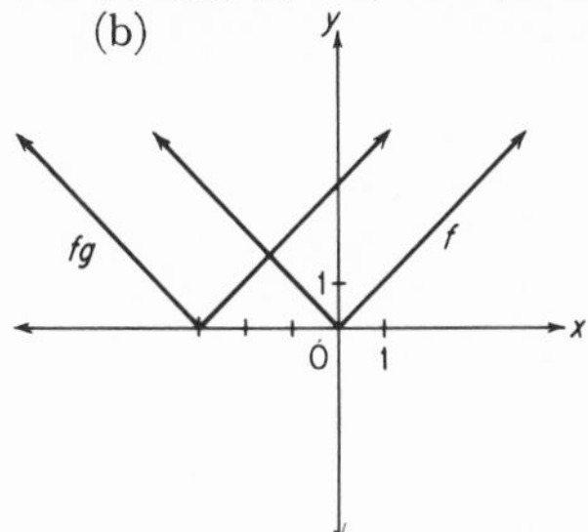

13. Translate (shift) the graph of f by c units (to the right or left) to obtain the graph of fg. The translation is to the right if c is negative, to the left if c is positive. The two graphs are congruent; translation is a "rigid motion."

6-4 Inverse Functions

1. (a) The numbers 3 and 1 have the same image. (b) Restrict the domain to $\{x: x \geqq 2\}$.

3. $\{(5, 1), (6, 2), (8, 3), (7, 4)\}$.

5. (a) 8; 0; (b) $g(x) = 2(x - 3)$.

7.

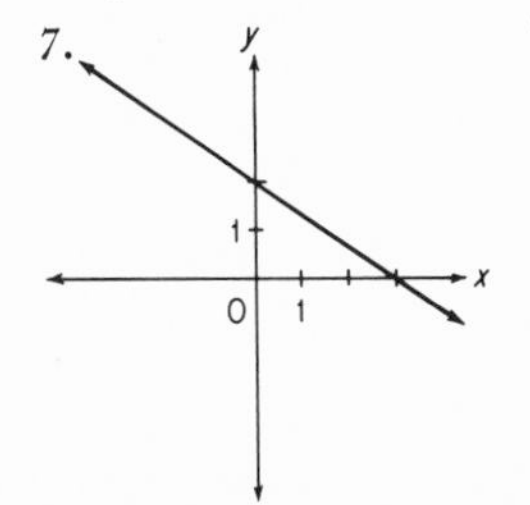

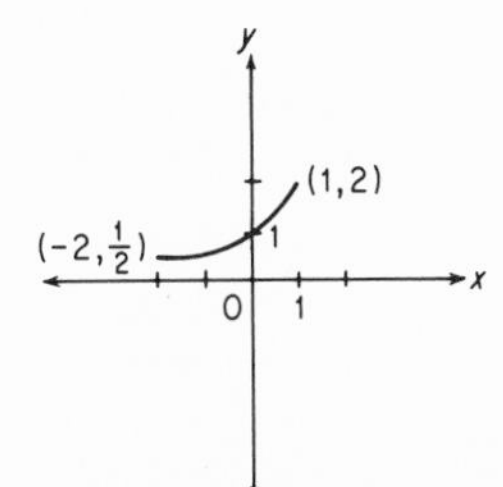

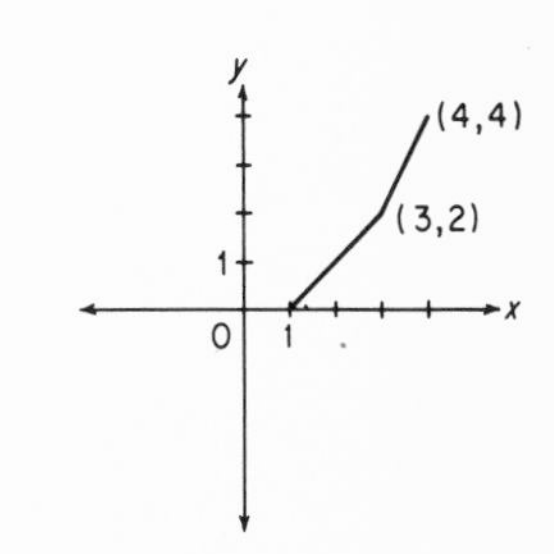

9. (a) Let g denote the inverse of f

x	$g(x)$
8	3
4	2
2	1
1	0
$\frac{1}{2}$	-1
$\frac{1}{4}$	-2
$\frac{1}{8}$	-3

(b)

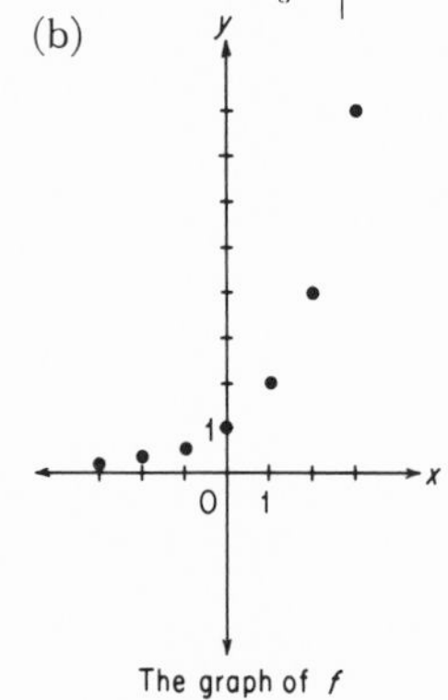

The graph of f

The graph of the inverse of f

(b)

11.

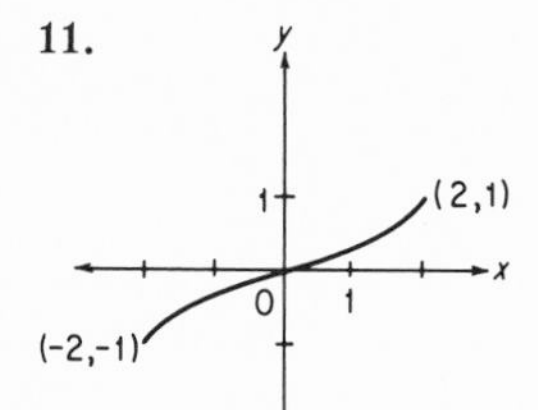

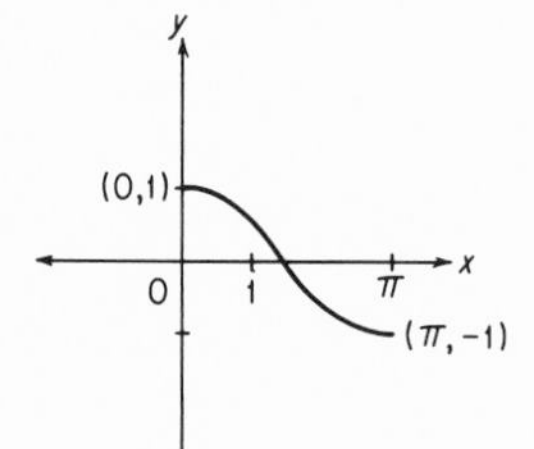

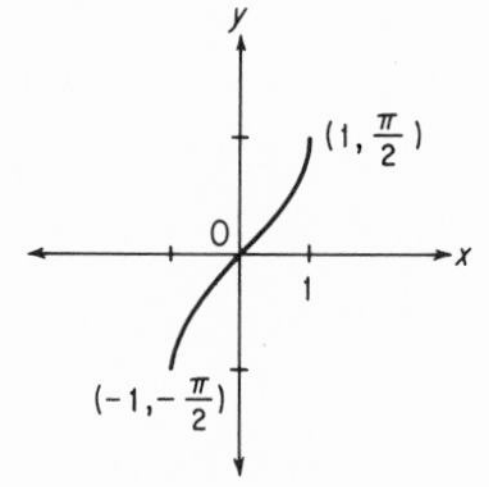

CHAPTER 7—EXPONENTIAL FUNCTIONS

7-1 Review of Integral Exponents

1. (a) 10^6; (b) 10^9; (c) 10^{10}; (d) 10^{-2}; (e) 10^{-4}; (f) 10^6; (g) 10^{20}; (h) 10^7; (i) 10^{-6}; (j) 10^4; (k) 10^{16}; (l) 10^{-12}; (m) 10^{-18}; (n) 10^0; (o) 10^{-4}; (p) 10^8; (q) 10^{16}; (r) 10^{-2}.

5. (a) $2^2 \cdot 5$; (b) $2^{-2} \cdot 5^{-1}$; (c) $2^2 \cdot 5^2$; (d) $2^{30} \cdot 5^{30}$; (e) $2^{-30} \cdot 5^{-30}$; (f) $2^4 \cdot 5$; (g) $2^{-4} \cdot 5^{-1}$; (h) $2^{40} \cdot 5^{40}$.

3. It is necessary to define $a^0 = 1$ if the properties of integral exponents (listed in Theorem 7-1) are to be valid without exception. In particular, $1 = a^n/a^n = a^{n-n} = a^0$.

7. (a) $\frac{3}{2}$; (b) $\frac{3}{2}$; (c) $\frac{1}{2}$; (d) $\frac{1}{3}$; (e) $\frac{3}{2}$; (f) $\frac{5}{6}$; (g) $-\frac{1}{3}$; (h) $-\frac{3}{2}$; (i) $\frac{5}{2}$; (j) $\frac{4}{5}$; (k) $\frac{1}{2}$; (l) $\frac{1}{3}$.

7-2 Scientific Notation

1. (a) 4.32×10^4; (b) 1×10^6; (c) 1.39×10^{-1}; (d) 3.5018×10^1; (e) 3.576×10^{-3}; (f) 8.4×10^{-8}; (g) 1.1×10^{16}; (h) 5.03×10^{14}; (i) 2.2×10^{24}; (j) 1.001×10^{24}; (k) 3.52×10^7; (l) 5.38×10^8.

3. (a) 1.25×2^3; (b) 1.3125×2^5; (c) 1.5625×2^5; (d) 1.5625×2^6; (e) 1.5625×2^7; (f) 1.6×2^{-1}.

7-3 Some Applications of Exponential Functions

1. (a)

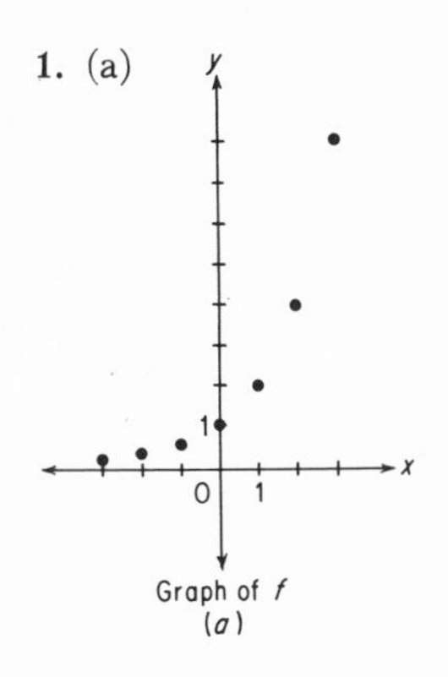

Graph of f
(a)

Graph of $-f$
(b)

(b) 8; (c) 4; (d) 1; (e) 16; (f) 2; (g) 8.

5. \$68,060 (to nearest ten dollars).

9. $N(4) = 50{,}625$.

3. (a) \$1469 (to nearest dollar); (b) \$2159 (to nearest dollar); (c) $\$1000 \times (1.08)^n$.

7. (a) $\frac{5}{16}$ ft; (b) $10 \times (\frac{1}{2})^n$ ft.

7-4 The Rational Numbers as a Subset of the Real Number System

1. (a) $0.1\bar{6}\ldots$; (b) $1.08\bar{3}\ldots$; (c) 0.875; (d) $0.\bar{1}\ldots$; (e) $0.\overline{142857}\ldots$; (f) $0.1\bar{3}\ldots$; (g) $1.1\bar{6}\ldots$; (h) 1.35.

3. (a) $\frac{69}{20}$; (b) $67/10{,}000$; (c) $\frac{2}{3}$; (d) $\frac{7}{9}$; (e) $\frac{31}{99}$; (f) $\frac{5}{11}$; (g) $\frac{1}{30}$; (h) $\frac{2}{15}$.

7-5 The Irrational Numbers and Their Relation to the Rational Numbers

1. (e) $\sqrt{3}$ is irrational (Theorem 7-5); (f) $\sqrt[4]{30}$ is irrational (Theorem 7-5).

3. (a) $(4 - \sqrt{2}) + \sqrt{2} = 4$; (b) $\sqrt{2}\cdot\sqrt{2} = 2$.

5. Let $r = 12.57$.

7. 2.828.

7-6 Exponential Functions that Have the Real Numbers as Their Domain

1.

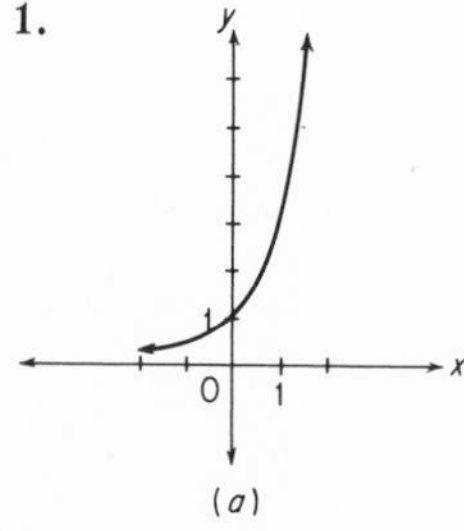

(a)

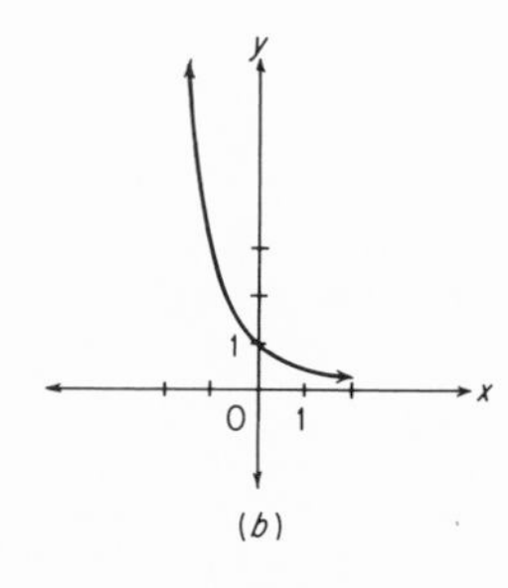

(b)

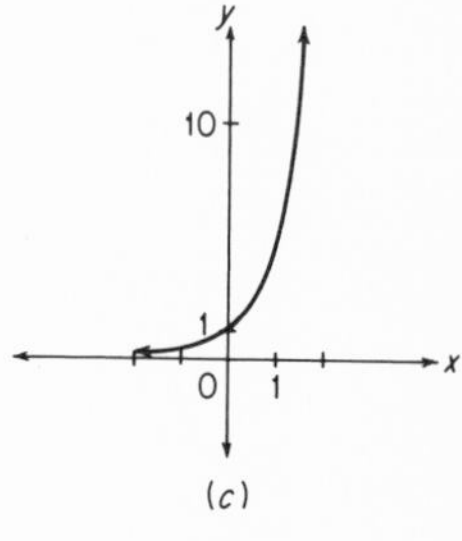

(c)

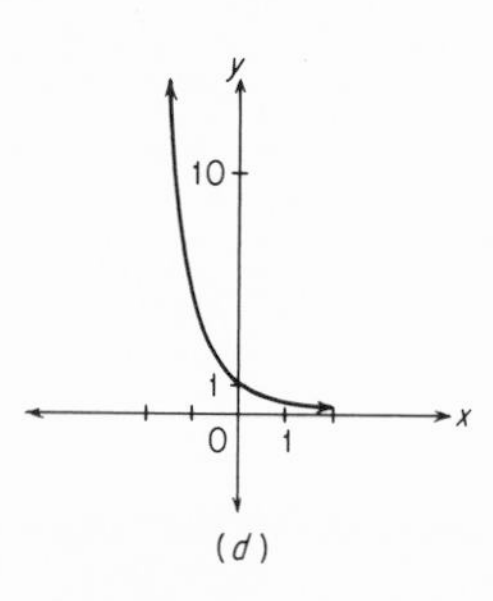

(d)

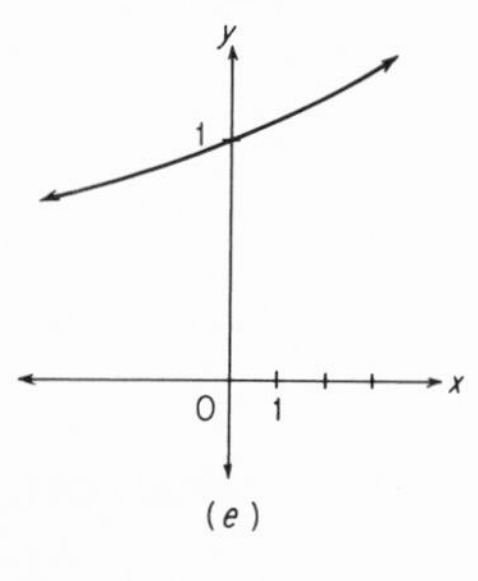

(e)

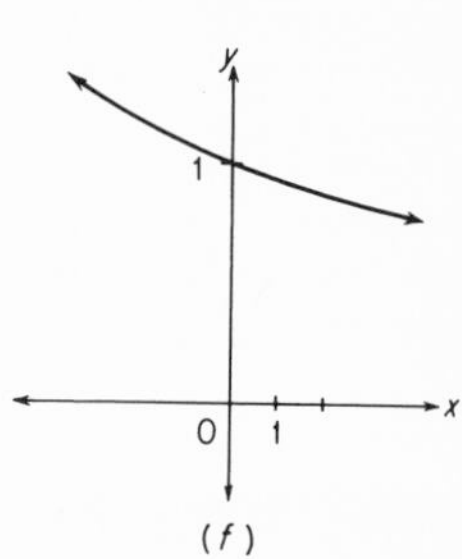

(f)

7-7 Rational Numbers as Exponents

1. (a) 4; (b) $\frac{1}{4}$; (c) 32; (d) 2; (e) 8; (f) $\frac{1}{8}$; (g) 100,000; (h) 0.00001; (i) 0.1; (j) 0.00001; (k) 100,000; (l) 100,000; (m) 8; (n) 3; (o) $\frac{1}{3}$; (p) 1000; (q) 100; (r) 100.

3. (a) 4; (b) 3; (c) 2; (d) 1; (e) 2; (f) $\frac{3}{2}$; (g) 2; (h) -3; (i) -3; (j) 1; (k) 6; (l) $-\frac{3}{2}$; (m) $\frac{2}{3}$; (n) 2.

7-8 The Function $F(x) = 10^x$, Domain: the Real Numbers

1. (a) $4 + 0.586$; (b) $-5 + 0.414$; (c) $-4 + 0.219$; (d) $-1 + 0.9939$; (e) $-11 + 0.02$; (f) $-2 + 0.988$.

3. (a) $\log 100 = 2$;
(b) $\log 1000 = 3$;
(c) $\log 1{,}000{,}000 = 6$;
(d) $\log 1 = 0$;
(e) $\log 1{,}000{,}000{,}000 = 9$;
(f) $\log 0.1 = -1$;
(g) $\log 0.01 = -2$;
(h) $\log 0.001 = -3$;
(i) $\log 10^{50} = 50$;
(j) $\log 10^{-40} = -40$.

7-9 Using Exponents to Simplify Calculations

1. 21.7

3. 10.9

5. 2.82×10^8

7. 25.5

9. 0.000105

7-10 Exponential Growth and Decay

1.

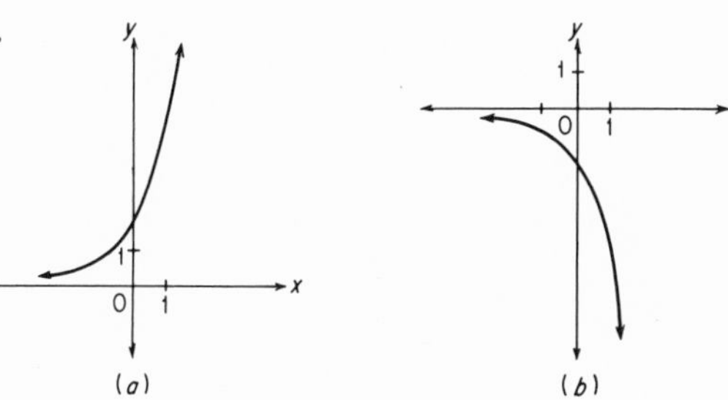

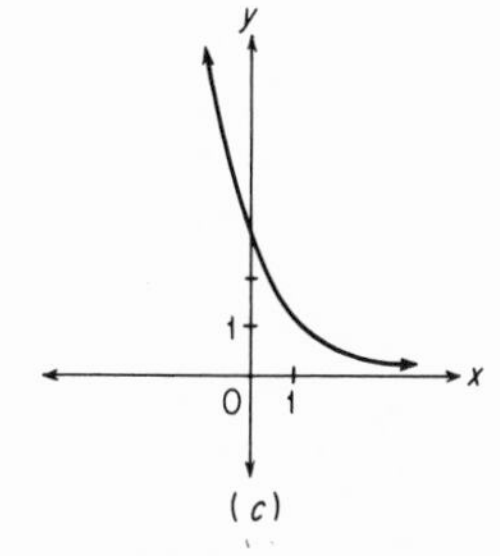

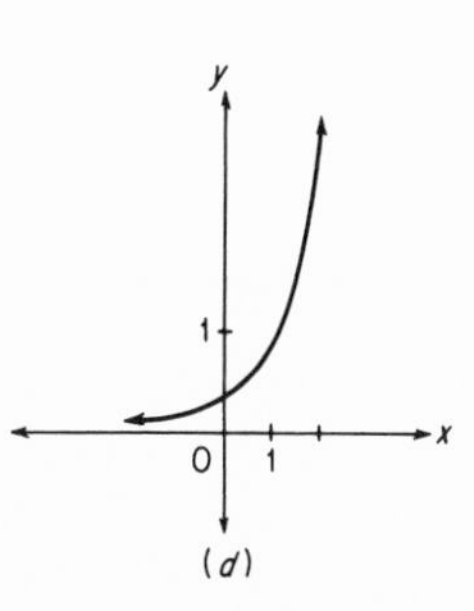

3. (a) $k = 4$, $a = \frac{1}{3}$; $g(x) = 4(\frac{1}{3})^x$; (b) $\frac{4}{81}$

5. (a) N_0; (b) $\frac{1}{2}$ unit of time.

7. (a) $\frac{256}{625} A_0$, where A_0 is the initial amount; (b) 3.1 years.

7-11 The Function $G(x) = \log x$, Domain: the Positive Numbers

1. (a) 1; (b) 2; (c) 3; (d) 6; (e) 0; (f) -2; (g) -3; (h) -1; (i) 40; (j) n.

3. (a) $2 < \log 256 < 3$.
(b) $3 < \log 1897 < 4$.
(c) $2 < \log 687 < 3$.
(d) $4 < \log 89{,}765 < 5$.
(e) $-1 < \log 0.89 < 0$.
(f) $-3 < \log 0.0089 < -2$.
(g) $-1 < \log 0.7 < 0$.
(h) $-5 < \log 0.00008 < -4$.

7-12 A Family of Logarithmic Functions

1.

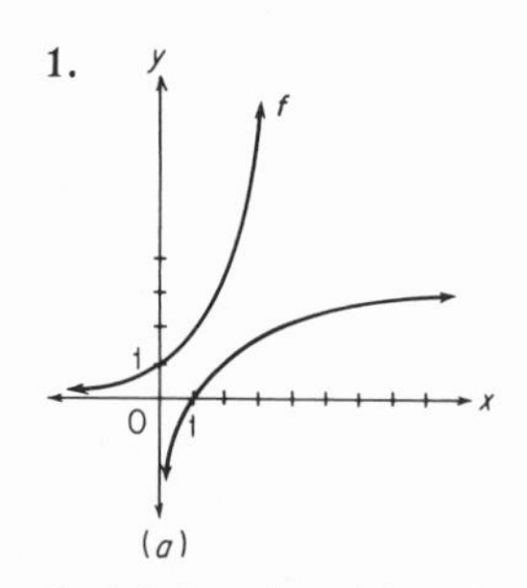

(*a*)

(*b*)

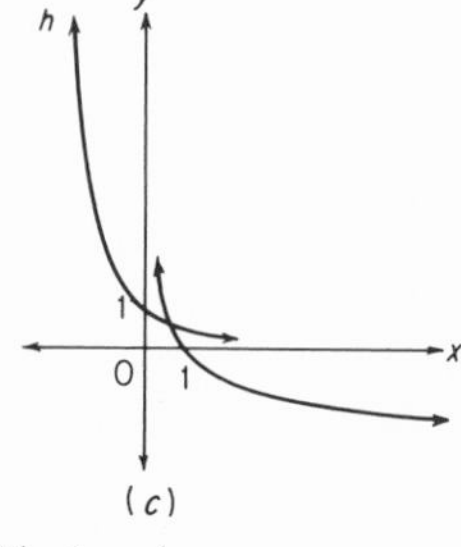

(*c*)

3. (a) $3 < \log_2 15 < 4$.
(b) $6 < \log_2 75 < 7$.
(c) $6 < \log_2 127 < 7$.
(d) $8 < \log_2 400 < 9$.
(e) $9 < \log_2 1000 < 10$.
(f) $-1 < \log_2 0.6 < 0$.
(g) $-5 < \log_2 0.06 < -4$.
(h) $2 < \log_3 19 < 3$.
(i) $-1 < \log_3 (\frac{1}{2}) < 0$.
(j) $3 < \log_3 45 < 4$.
(k) $2 < \log_5 79 < 3$.
(l) $-6 < \log_{1/2} 43 < -5$.

CHAPTER 8—PERIODIC FUNCTIONS

8-1 What Is a Periodic Function?

1. (a) 0; (b) 1; (c) 1; (d) 1; (e) 0; (f) 1.

3. (a) Saturday; (b) Friday; (c) Saturday.

5. (a) 2 P.M.; (b) 2 A.M.

7. (a)

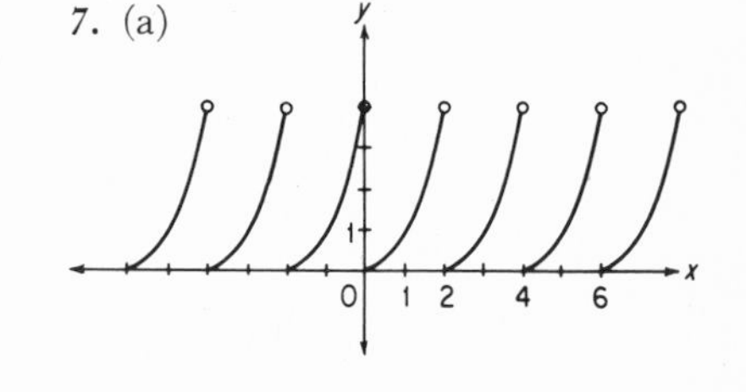

(b) 1; (c) 0; (d) 1; (e) $\frac{9}{4}$.

9. (a)

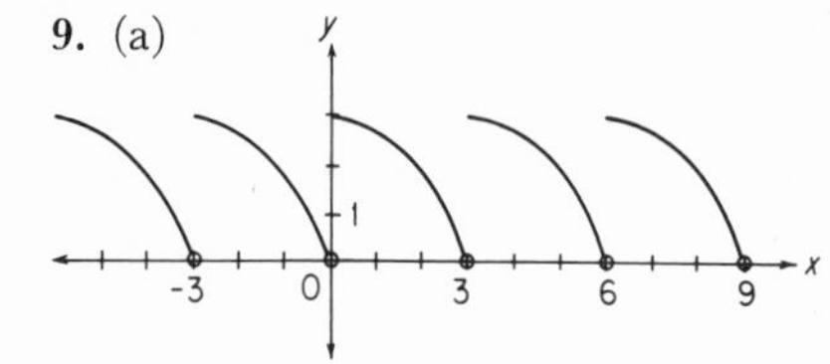

(b) 3; (c) $\sqrt{5}$.

13. (a) Yes. Suppose $n \leqq x < n + 1$, where n and $n + 1$ denote consecutive integers. Then $n + 1 \leqq x + 1 < n + 2$. So $[x] = n$ and $[x + 1] = n + 1$. Therefore, $[x + 1] = 1 + [x]$.
(b) $f(x + 1) = x + 1 - [x + 1] = x + 1 - ([x] + 1) = x - [x]$. To show that a number p is not a period if $0 < p < 1$, show that $f[(1 - p/2) + p] \neq f(1 - p/2)$.
(c)

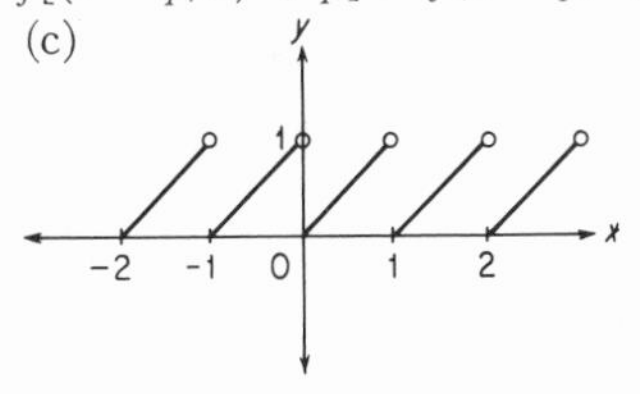

11. The domain of K is the real numbers; K is periodic with period 2; $K(x) = (x - 1)^2$ on the interval $\{x: 0 \leqq x < 2\}$.

15. No. Suppose f is periodic with period p. Then $f(p) = f(0 + p) = f(0)$. This contradicts the assumption that f is a one-to-one function.

8-2 Some Square Functions

1.

t	$P(t)$	$C(t)$	$S(t)$
10	(0, 1)	0	1
15	(1, −1)	1	−1
17	(1, 1)	1	1
18	(0, 1)	0	1
20	(−1, 0)	−1	0
100	(−1, 0)	−1	0
1000	(1, 0)	1	0
−80	(1, 0)	1	0
−14	(0, 1)	0	1
−1000	(1, 0)	1	0
10^6	(1, 0)	1	0

5.

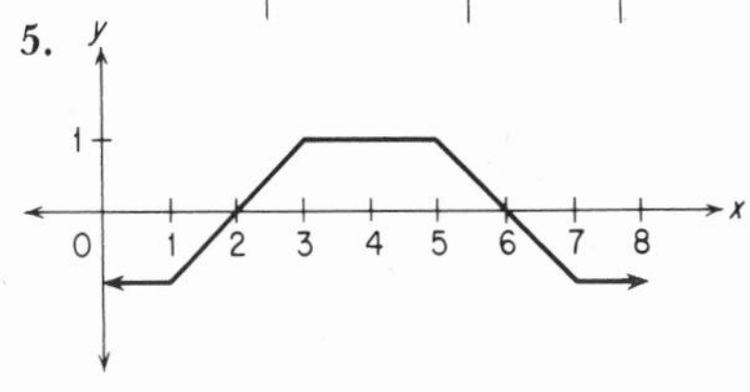

3. (a)

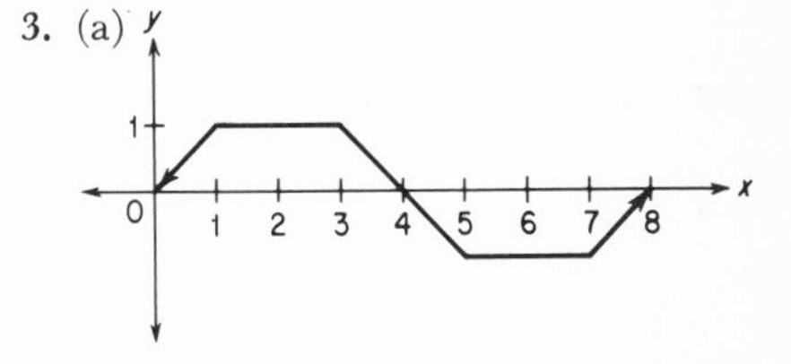

(b) Yes. Translate (shift) the graph of C two units to the right to obtain the graph of S.

7. (a) Valid. Observe that the multipart rule for C involves only rational numbers and the subtraction of a number from a rational number. The difference of two rational numbers is a rational number. (b) Invalid. For example, $C(\sqrt{2}/2) = 1$.

8-3 The Sine and Cosine Functions

1. Since the range of the function Q is $\{(x, y): x^2 + y^2 = 1\}$, then the range of the cosine function is all numbers x such that $x^2 + y^2 = 1$ for some number y. Now $x^2 + y^2 = 1 \Rightarrow x^2 \leqq 1 \Rightarrow -1 \leqq x \leqq 1$. Furthermore, if $-1 \leqq x \leqq 1$ then x is in the range of the cosine function because for such a number there is a number $y = \sqrt{1 - x^2}$ such that $x^2 + y^2 = 1$. A similar argument can be made about the range of the sine function.

3.

t	$Q(t)$	$\cos t$	$\sin t$
$\frac{3\pi}{2}$	$(0, -1)$	0	-1
$\frac{-3\pi}{2}$	$(0, 1)$	0	1
5π	$(-1, 0)$	-1	0
-5π	$(-1, 0)$	-1	0
6π	$(1, 0)$	1	0
-6π	$(1, 0)$	1	0
$\frac{7\pi}{2}$	$(0, -1)$	0	-1
$\frac{-7\pi}{2}$	$(0, 1)$	0	1
8π	$(1, 0)$	1	0
-8π	$(1, 0)$	1	0
1000π	$(1, 0)$	1	0

5.

t	$\cos t$	$\sin t$
$\frac{3\pi}{4}$	$\frac{-\sqrt{2}}{2}$	$\frac{\sqrt{2}}{2}$
$\frac{5\pi}{4}$	$\frac{-\sqrt{2}}{2}$	$\frac{-\sqrt{2}}{2}$
$\frac{7\pi}{4}$	$\frac{\sqrt{2}}{2}$	$\frac{-\sqrt{2}}{2}$
$\frac{37\pi}{4}$	$\frac{-\sqrt{2}}{2}$	$\frac{-\sqrt{2}}{2}$
$\frac{-\pi}{4}$	$\frac{\sqrt{2}}{2}$	$\frac{-\sqrt{2}}{2}$
$\frac{-7\pi}{4}$	$\frac{\sqrt{2}}{2}$	$\frac{\sqrt{2}}{2}$
$\frac{-9\pi}{4}$	$\frac{\sqrt{2}}{2}$	$\frac{-\sqrt{2}}{2}$

7. (a) $\{y: -2 \leqq y \leqq 2\}$.
(b) $\{y: -1 \leqq y \leqq 1\}$.
(c) $\{y: 0 \leqq y \leqq 1\}$.
(d) $\{y: -1 \leqq y \leqq 1\}$.
(e) $\{y: -1 \leqq y \leqq 1\}$.
(f) $\{y: -1 \leqq y \leqq 1\}$.
(g) The real numbers.
(h) The real numbers.

9.

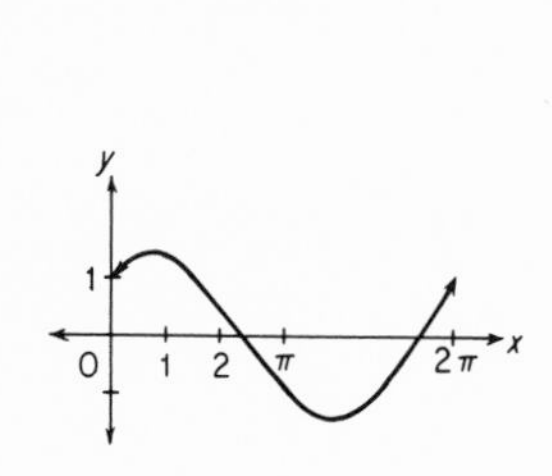

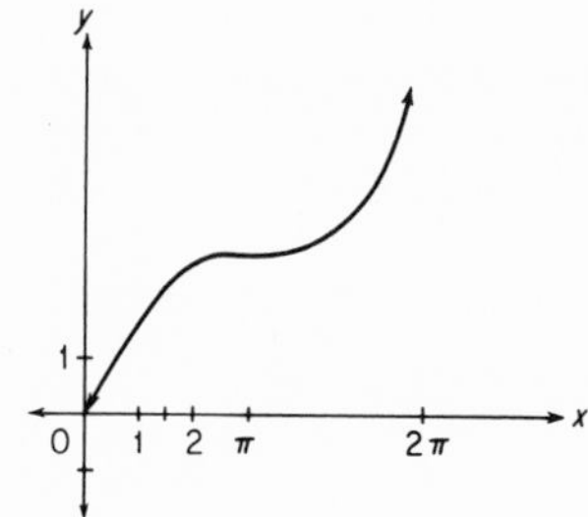

8-4 The Addition Formulas

1. $\cos\left(\frac{\pi}{2}+\frac{\pi}{2}\right) = \cos\pi = -1$, but $\cos\frac{\pi}{2}+\cos\frac{\pi}{2} = 0$.

3. (a) $\sin\left(\theta+\frac{\pi}{2}\right) = \sin\theta\cos\frac{\pi}{2} + \cos\theta\sin\frac{\pi}{2} = (\sin\theta)\cdot 0 + (\cos\theta)\cdot 1 = \cos\theta$.

   (b) $\cos\left(\theta+\frac{\pi}{2}\right) = \cos\theta\cos\frac{\pi}{2} - \sin\theta\sin\frac{\pi}{2} = (\cos\theta)\cdot 0 - (\sin\theta)\cdot 1 = -\sin\theta$.

   (c) $\sin\left(\frac{\pi}{2}-\theta\right) = \sin\frac{\pi}{2}\cos\theta - \sin\theta\cos\frac{\pi}{2} = 1\cdot\cos\theta - (\sin\theta)\cdot 0 = \cos\theta$.

   (d) $\cos\left(\frac{\pi}{2}-\theta\right) = \cos\frac{\pi}{2}\cos\theta + \sin\frac{\pi}{2}\sin\theta = 0\cdot\cos\theta + 1\cdot\sin\theta = \sin\theta$.

5. $\sin(2x) = \sin(x+x) = \sin x\cos x + \sin x\cos x = 2\sin x\cos x$.

7. (a)
$$\begin{aligned}\cos(3x) &= \cos(2x+x)\\ &= \cos(2x)\cos x - \sin(2x)\sin x\\ &= (2\cos^2 x - 1)\cos x - (2\sin x\cos x)\sin x\\ &= 2\cos^3 x - \cos x - 2\cos x(1-\cos^2 x)\\ &= 4\cos^3 x - 3\cos x.\end{aligned}$$

   (b)
$$\begin{aligned}\sin(3x) &= \sin(2x+x)\\ &= \sin(2x)\cos x + \sin x\cos(2x)\\ &= 2\sin x\cos^2 x + \sin x(1-2\sin^2 x)\\ &= 2\sin x(1-\sin^2 x) + \sin x(1-2\sin^2 x)\\ &= -4\sin^3 x + 3\sin x.\end{aligned}$$

8-5 More on the Sine and Cosine Functions

1. (a) $2.45 \approx 1(\pi/2) + 0.88$.
   (b) $5.74 \approx 3(\pi/2) + 1.03$.
   (c) $30 \approx 19(\pi/2) + 0.15$.
   (d) $7.16 \approx 4(\pi/2) + 0.88$.
   (e) $1.8 \approx 1(\pi/2) + 0.23$.
   (f) $4.73 \approx 3(\pi/2) + 0.02$.
   (g) $3 \approx 1(\pi/2) + 1.43$.
   (h) $3\pi = 6(\pi/2) + 0$.

3. (a) 0.0208; (b) 0.3820; (c) 0.9316; (d) -0.8975; (e) 0.1925; (f) 0.8283.

8-6 Functions Defined by $f(x) = A \sin (Bx + C)$

1.

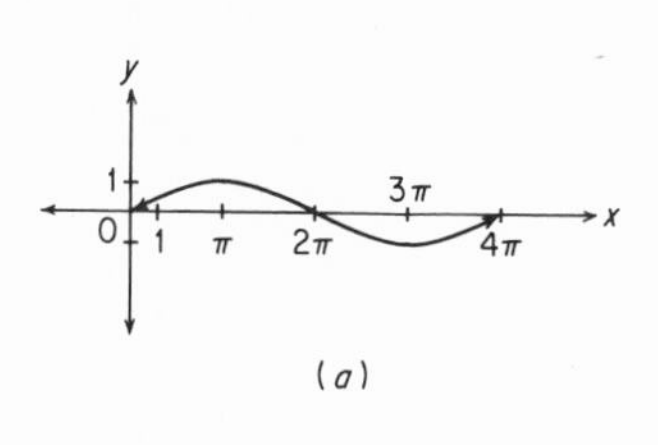

(a)

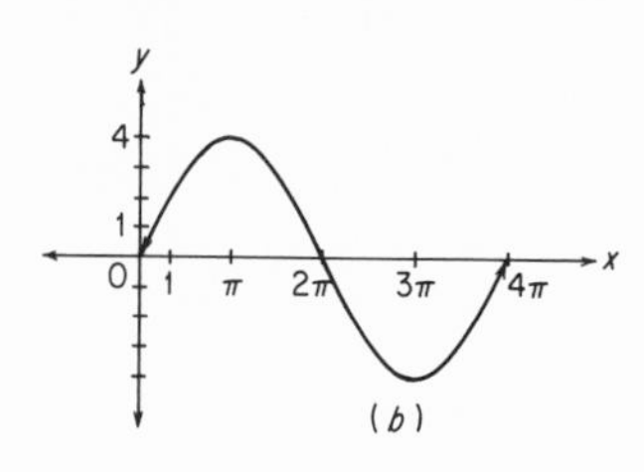

(b)

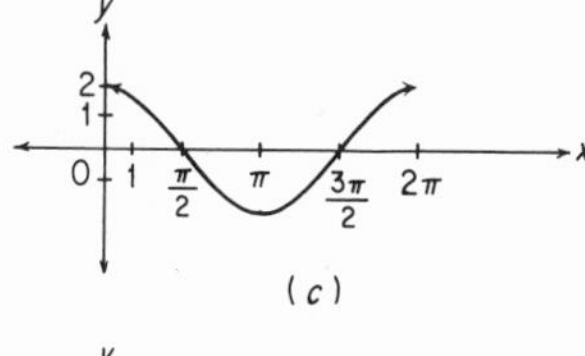

(c)

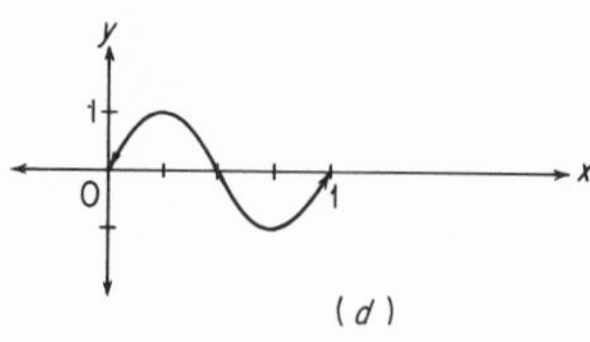

(d)

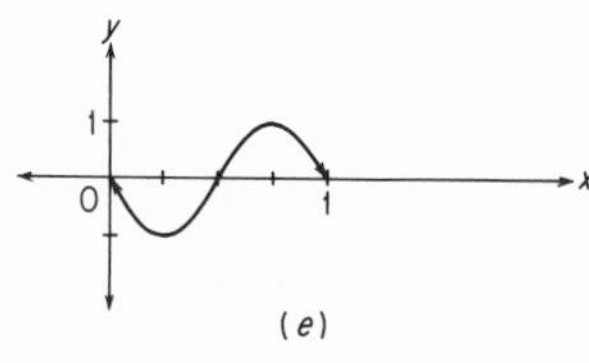

(e)

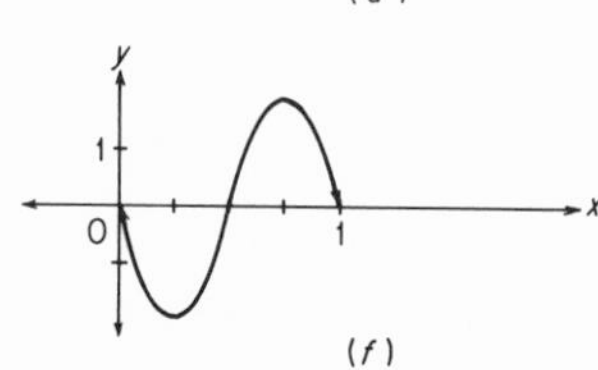

(f)

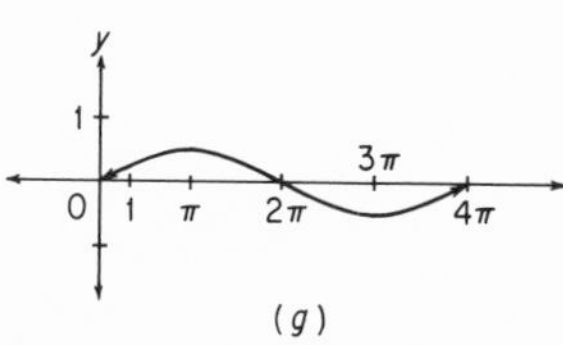

(g)

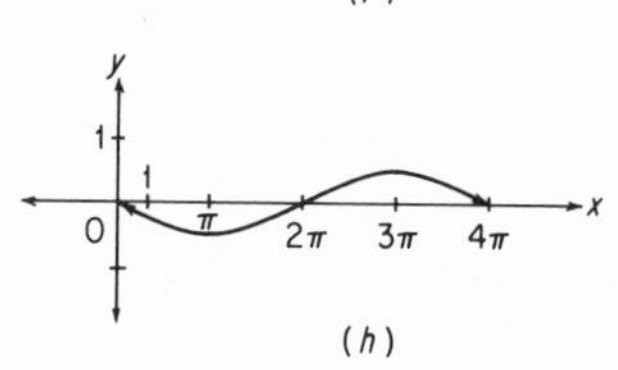

(h)

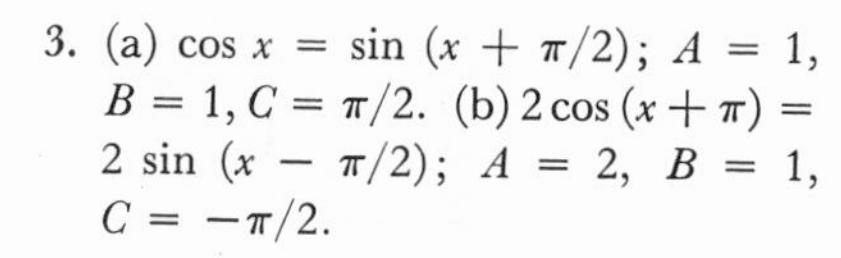
3. (a) $\cos x = \sin (x + \pi/2)$; $A = 1$, $B = 1, C = \pi/2$. (b) $2 \cos (x + \pi) = 2 \sin (x - \pi/2)$; $A = 2$, $B = 1$, $C = -\pi/2$.

5.

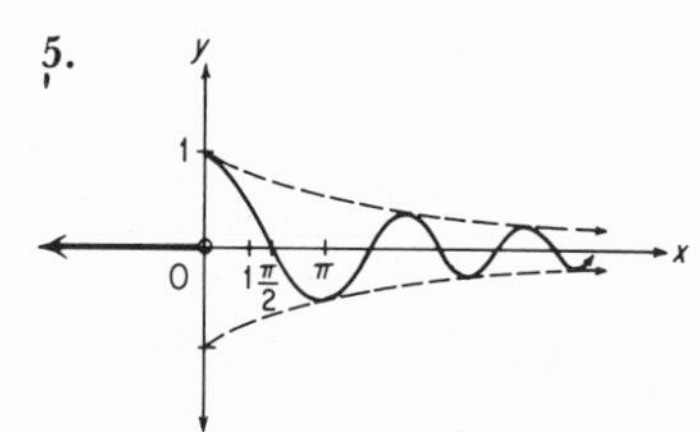

8-7 Functions Constructed from the Sine and Cosine Functions

1. (a) 2; (b) 2; (c) 0; (d) 1; (e) 1; (f) Not defined; (g) $\sqrt{3}$; (h) $-\sqrt{3}$; (i) -2; (j) -1; (k) $2\sqrt{3}/3$; (l) $\sqrt{2}$; (m) -2; (n) 1; (o) -1; (p) 0; (q) $\sqrt{3}/3$; (r) -1.

3. $\tan (-x) = \dfrac{\sin (-x)}{\cos (-x)} = \dfrac{-\sin x}{\cos x} = -\tan x.$

5. $\csc (-x) = \dfrac{1}{\sin (-x)} = \dfrac{1}{-\sin x} = -\csc x.$

7.

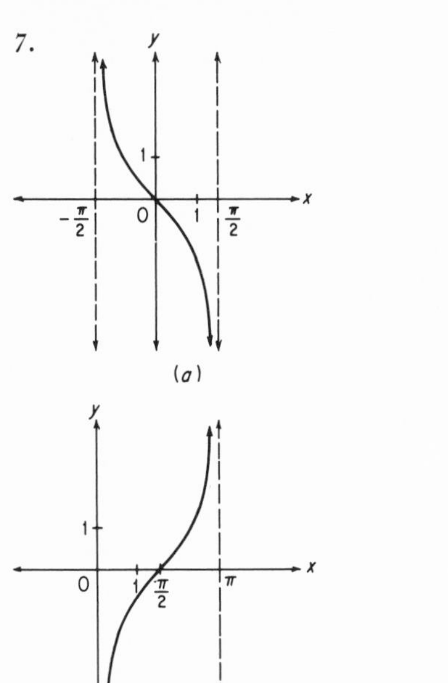

(a) (b)

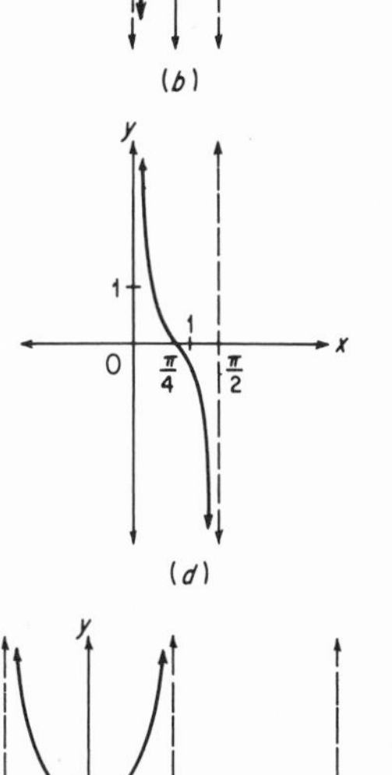

(c) (d)

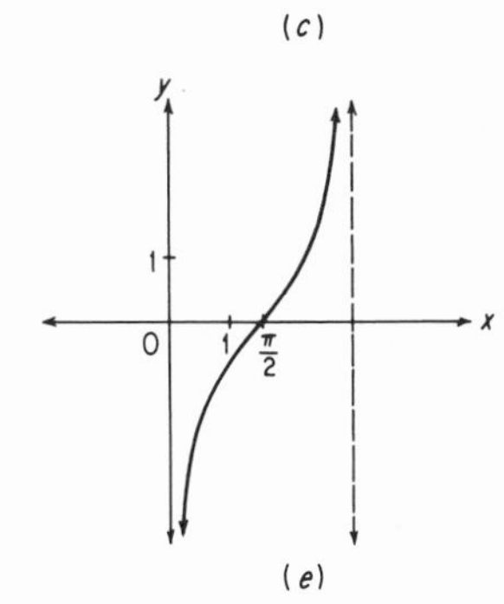

(e)

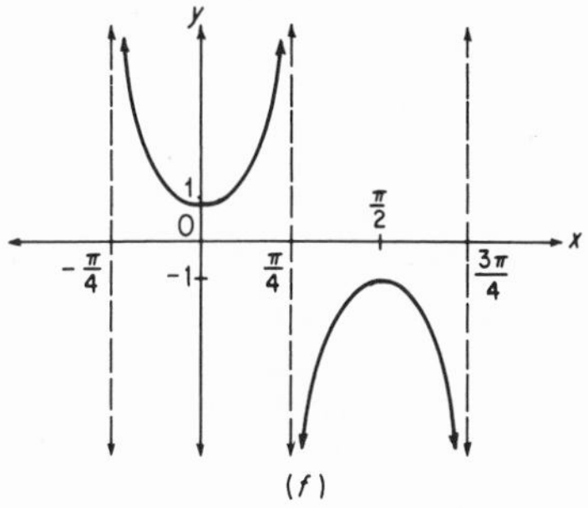

(f)

9. $\tan(x+y) = \dfrac{\sin(x+y)}{\cos(x+y)}$

$$= \frac{\dfrac{\sin x \cos y}{\cos x \cos y} + \dfrac{\cos x \sin y}{\cos x \cos y}}{\dfrac{\cos x \cos y}{\cos x \cos y} - \dfrac{\sin x \sin y}{\cos x \cos y}}$$

$$= \frac{\tan x + \tan y}{1 - \tan x \tan y}.$$

11. $\tan^2 x + 1 = \dfrac{\sin^2 x}{\cos^2 x} + 1$

$$= \frac{\sin^2 x + \cos^2 x}{\cos^2 x} = \sec^2 x.$$

13. $\tan(x+\pi) = \dfrac{\tan x + \tan \pi}{1 - \tan x \tan \pi}$

$$= \tan x.$$

8-8 Does f Map any Number onto c?

1. (a) $2\pi/3$; (b) None; (c) 0; (d) 0.48; (e) $\pi/4$; (f) $\pi/6$; (g) $\pi/2$; (h) 2.19.

3. (a) $-\pi/4$; (b) 1.19; (c) $\pi/3$; (d) 0.41; (e) 0.46; (f) 0.90; (g) $\pi/4$; (h) 0.32.

5. (a) $\{x: x = \pi/6 + 2n\pi\} \cup \{x: x = 5\pi/6 + 2n\pi\}$.
(b) $\{x: x = \pi/4 + n\pi\}$.
(c) $\{x: x = 2n\pi\}$.
(d) $\{x: x = 0.16 + 2n\pi\} \cup \{x: x = 2.98 + 2n\pi\}$.

8-9 The Arccosine, Arcsine, and Arctangent Functions

1. (a) $-\pi/2$; (b) $\pi/3$; (c) $2\pi/3$; (d) $-\pi/4$; (e) 0; (f) $\pi/2$; (g) 0; (h) 0; (i) $\frac{1}{3}$; (j) 4.

3. (a) Arcsin 0.4; (b) Arctan 47; (c) Arccos (-0.6); (d) Arctan $\frac{1}{3}$.

5. Let $f(x)$ = cotangent x, domain: $\{x: 0 < x < \pi\}$. The inverse of this function is called the Arccotangent function. Its domain is the set of real numbers.

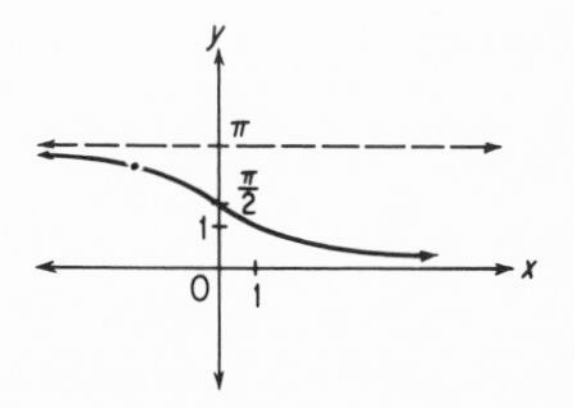

8-10 How to Measure an Angle

1. (a) $(360/\pi)°$; (b) $(540/\pi)°$; (c) $(270/\pi)°$; (d) 30°; (e) 60°; (f) 270°.

3. (a) $\pi/4$; (b) $3\pi/4$; (c) $\pi/90$; (d) $2\pi/3$; (e) $5\pi/6$; (f) $\pi/18$.

5. (a) To "measure" an angle in this way it would be necessary only to measure the length of a line segment, which is a familiar idea. (b) The "line segment measure" of $\angle AVC$ (adjoining figure) does not equal the line segment measure of $\angle AVB$ plus the line segment measure of $\angle BVC$.

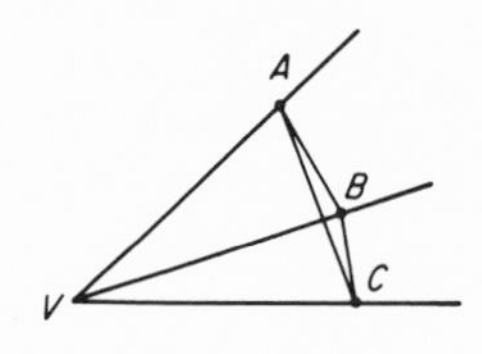

8-11 The Trigonometry of Right Triangles

1. (a) 0.39; 1.2; (b) 23°; 67°.

3. 31°; 59°.

5. 173 ft.

7. (a) 60°; (b) 30°.

9. 600 ft.

11. 47.7 ft.

13. (a) 4.00×10^4 ft; (b) 45°.

Index